Flash Memory Devices

Flash Memory Devices

Editors

Cristian Zambelli
Rino Micheloni

MDPI • Basel • Beijing • Wuhan • Barcelona • Belgrade • Manchester • Tokyo • Cluj • Tianjin

Editors

Cristian Zambelli
Dipartimento di Ingegneria,
Università di Ferrara
Italy

Rino Micheloni
Dipartimento di Ingegneria,
Università di Ferrara
Italy

Editorial Office
MDPI
St. Alban-Anlage 66
4052 Basel, Switzerland

This is a reprint of articles from the Special Issue published online in the open access journal *Micromachines* (ISSN 2072-666X) (available at: http://www.mdpi.com).

For citation purposes, cite each article independently as indicated on the article page online and as indicated below:

LastName, A.A.; LastName, B.B.; LastName, C.C. Article Title. *Journal Name* **Year**, *Volume Number*, Page Range.

ISBN 978-3-0365-3012-3 (Hbk)
ISBN 978-3-0365-3013-0 (PDF)

Contents

About the Editors . . . . **vii**

Cristian Zambelli and Rino Micheloni
Editorial for the Special Issue on Flash Memory Devices
Reprinted from: *Micromachines* **2021**, *12*, 1566, doi:10.3390/mi12121566 . . . . **1**

Young Suh Song and Byung-Gook Park
Retention Enhancement in Low Power NOR Flash Array with High-κ–Based Charge-Trapping Memory by Utilizing High Permittivity and High Bandgap of Aluminum Oxide
Reprinted from: *Micromachines* **2021**, *12*, 328, doi:10.3390/mi12030328 . . . . **5**

Su-in Yi and Jungsik Kim
Novel Program Scheme of Vertical NAND Flash Memory for Reduction of Z-Interference
Reprinted from: *Micromachines* **2021**, *12*, 584, doi:10.3390/mi12050584 . . . . **17**

Alessandro S. Spinelli, Gerardo Malavena, Andrea L. Lacaita and Christian Monzio Compagnoni
Random Telegraph Noise in 3D NAND Flash Memories
Reprinted from: *Micromachines* **2021**, *12*, 703, doi:10.3390/mi12060703 . . . . **27**

Haichun Zhang, Jie Wang, Zhuo Chen, Yuqian Pan, Zhaojun Lu and Zhenglin Liu
An SVM-Based NAND Flash Endurance Prediction Method
Reprinted from: *Micromachines* **2021**, *12*, 746, doi:10.3390/mi12070746 . . . . **41**

Michele Favalli, Cristian Zambelli, Alessia Marelli, Rino Micheloni and Piero Olivo
A Scalable Bidimensional Randomization Scheme for TLC 3D NAND Flash Memories
Reprinted from: *Micromachines* **2021**, *12*, 759, doi:10.3390/mi12070759 . . . . **63**

Yajuan Du, Wei Liu, Yuan Gao, Rachata Ausavarungnirun
Observation and Optimization on Garbage Collection of Flash Memories: The View in Performance Cliff
Reprinted from: *Micromachines* **2021**, *12*, 846, doi:10.3390/mi12070846 . . . . **77**

Ruiquan He, Haihua Hu; Chunru Xiong; Guojun Han
Artificial Neural Network Assisted Error Correction for MLC NAND Flash Memory
Reprinted from: *Micromachines* **2021**, *12*, 879, doi:10.3390/mi12080879 . . . . **95**

Sivaramakrishnan Ramesh, Arjun Ajaykumar, Lars-Åke Ragnarsson, Laurent Breuil, Gabriel Khalil El Hajjam, Ben Kaczer, Attilio Belmonte, Laura Nyns, Jean-Philippe Soulié, Geert Van den bosch and Maarten Rosmeulen
Understanding the Origin of Metal Gate Work Function Shift and Its Impact on Erase Performance in 3D NAND Flash Memories
Reprinted from: *Micromachines* **2021**, *12*, 1084, doi:10.3390/mi12091084 . . . . **107**

Fei Chen, Bo Chen, Hongzhe Lin, Yachen Kong, Xin Liu, Xuepeng Zhan and Jiezhi Chen
Temperature Impacts on Endurance and Read Disturbs in Charge-Trap 3D NAND Flash Memories
Reprinted from: *Micromachines* **2021**, *12*, 1152, doi:10.3390/mi12101152 . . . . **123**

About the Editors

Cristian Zambelli received M.Sc. and the Ph.D. degrees in Electronic Engineering from the University of Ferrara, Ferrara, Italy, in 2008 and 2012, respectively. Since 2015, he has held an Assistant Professor position with the same institution. His current research interests include the electrical characterization, physics, and reliability modeling of different nonvolatile memories such as NAND/NOR Flash, Phase Change Memories, Nano-MEMS memories, Resistive RAM (RRAM), and Magnetic RAM. He is also interested in the evaluation of the Solid State Drive reliability/performance trade-offs exposed by the integrated memory technology.

Rino Micheloni is a Research Fellow at the University of Ferrara, Italy. Before that, he was Vice-President and Fellow at PMC/Microsemi/Microchip, where he established the Flash Signal Processing Labs in Milan, Italy, with special focus on NAND Flash technology characterization, Machine Learning techniques for improving memory reliability, and Error Correction Codes. Before that, he was with IDT Inc. as Lead Flash Technologist, driving the architecture and design of the BCH engine in the world's first PCIe NVMe SSD controller. Early in his career, he led NAND design teams at STMicroelectronics, Hynix, and Infineon; during this time, he developed the industry's first MLC NOR device with embedded ECC and the industry's first MLC NAND with embedded BCH. Dr. Micheloni is an IEEE Senior Member, he has co-authored 100+ publications and 10 books on non-volatile memories, and he holds 295 patents. In 2020 Dr. Micheloni was selected for the European Inventor Award.

 micromachines

MDPI

Editorial

Editorial for the Special Issue on Flash Memory Devices

Cristian Zambelli * and Rino Micheloni

Dipartimento di Ingegneria, Università degli Studi di Ferrara, Via G. Saragat 1, 44122 Ferrara, Italy; rino.micheloni@ieee.org
* Correspondence: cristian.zambelli@unife.it; Tel.: +39-0532-974993

Citation: Zambelli, C.; Micheloni, R. Editorial for the Special Issue on Flash Memory Devices. *Micromachines* **2021**, *12*, 1566. https://doi.org/10.3390/mi12121566

Received: 9 December 2021
Accepted: 15 December 2021
Published: 17 December 2021

Publisher's Note: MDPI stays neutral with regard to jurisdictional claims in published maps and institutional affiliations.

Flash memory devices represented a breakthrough in the storage industry since their inception in the mid-1980s, and innovation is still ongoing after more than 35 years. They are the largest landscape of storage devices, and we expect more and more advancements in the coming years. The peculiarity of such technology is an inherent flexibility in terms of performance and integration density according to the architecture devised for integration of cells. In the context of code storage applications in the embedded world, automotive microcontrollers, IoT smart devices, and edge AI, we rely on NOR Flash technology. Their density ranges from a few Kbytes up to the Gigabit size. However, when massive data storage is required, NAND Flash memories are a must in a system. NAND Flash can be found in USB and Flash Cards (SD, eMMC), but most of all in Solid-State Drives (SSDs). Since SSDs are extremely demanding in terms of storage capacity, they fueled a new wave of innovation for Flash memories, namely 3D architecture. Today, 3D means that multiple layers (up to almost two hundred, as we speak) of memory cells are manufactured within the same piece of silicon, easily reaching a terabit of storage capacity per chip. This will require a lot of innovation in process technology, materials, circuit design, flash management algorithms, Error Correction Code (ECC), and finally system co-design for new applications such AI and security enforcement.

This Special Issue provides insight on and advancements in Flash memory devices. There are nine papers including one review paper, covering the reliability of 3D NAND Flash devices [1–3], the characterization and design of Flash memory cell/string [2,4,5], NOR Flash memories for embedded applications [5], a set of Error Correction Codes and Secondary Correction Algorithms for flash memories [6,7], Flash management through flash signal processing in controllers for Big Data storage [6–8], and the impact of Flash memories on Solid State Drives reliability and performance [8,9].

Flash memory devices integrated either with planar or 3D process scheme suffer from major performance and reliability threats that can be handled starting from the first manufacturing process steps. In [1], Spinelli et al. reviewed the phenomenology of random telegraph noise (RTN) in 3D NAND Flash arrays to deeply understand such a time-dependent reliability issue. They pointed out the relevant role played by the polycrystalline nature of the string silicon channels through experimental data and simulation models of the current transport. The RTN features changed significantly in the transition from planar to 3D processes due to the presence of highly defective grain boundaries on percolative current transport in cell channels in combination with the localized nature of the RTN traps. In [2], Ramesh et al. studied the erase operation performance by characterizing the metal gate work function of different metal electrode and high-k dielectric combinations in 3D Flash cell stack integration. They investigated the impact of different thermal treatments on the work function and observed a dipole formation at the metal/high-k and/or high-k/SiO_2 interfaces. They also concluded that the erase performance of metal/high-k/ONO/Si (MHONOS) capacitors is identical to the gate stack in three-dimensional (3D) NAND Flash, although the work function extraction is convoluted by the dipole formation. In [3], Chen et al. investigated the temperature effects that affect the reliability and performance of NAND flash memories. They characterized Triple-Level Cell (TLC) 3D NAND flash memory chips in a wide temperature range by focusing on the raw bit error rate (RBER) degradation during frequent-write (endurance) and frequent-read (read disturb) working conditions. It was observed that the program time shows strong dependence on the temperature and lifetime degradation induced by cycling and that the RBER can be suppressed at higher temperatures. Read disturb has been found to be more detrimental at low temperatures, but it can be beneficial for RBER recovery at high temperatures.

A successful Flash technology requires a careful design of the cell structure and of the operation modes. In [4], Yi et al. addressed the minimization of the threshold voltage variation of programmed cells by developing a new programming scheme to write the cells from the top array in vertical NAND (VNAND) structures to reach 5 bits per cell storage paradigm. With the aid of Technology-Computer-Aided Design (TCAD), the Z-Interference for this new program algorithm is found to be better than the state of the art by at least 20 mV. Moreover, under scaled cell dimensions, the improvement becomes protruding. In [5], Song et al. incorporated aluminum oxide in tunnel oxide to improve retention characteristics of NOR flash arrays. By adopting the proposed tunneling layers in the NOR flash array, the threshold voltage window after 10 years from programming and erasing (P/E) was improved by 4 V. The validation of the proposed device structure took place by comparing it with another stacked-engineered structure with $SiO_2/Si_3N_4/SiO_2$ tunneling layers. Simulations through TCAD were exploited in this context. In addition, to verify that our proposed structure is suitable for NOR flash array, disturbance issues are also carefully investigated.

As Flash architectures scale, their reliability worsens significantly and they require proactive control by using either advanced Error Correction Codes or some secondary correction mechanisms that help the recovery of the corrupted stored information. NAND flash memories are addressed especially in this context. In [6], Zhang et al. proposed a set of machine learning algorithms to accurately predict endurance levels of the memory array, which is of great significance for effectively extending the lifetime of NAND flash memory devices and avoiding serious losses caused by sudden failures. In this work, a multi-class endurance prediction scheme based on the SVM algorithm is proposed, which can predict the remaining endurance level and the RBER at various lifetime points. Feature analysis based on endurance data is used to determine the basic elements of the model and its implementation on a System-on-Chip (SoC) module showing the completion of a single prediction within 37 μs. In [7], He et al. presents a novel neural-network-assisted error correction (ANNAEC) scheme to increase the reliability of multi-level cell (MLC) NAND Flash memory. They propose a relative log-likelihood ratio (LLR) to estimate the actual LLR and transform the bit detection into a clustering problem suitable for a neural network to learn the error characteristics of the NAND flash memory channel. Simulation results show that the proposed scheme can significantly increase the lifetime of NAND flash memories.

The interaction of Flash memories at the system level as currently happens in Solid State Drive (SSD) architectures is also of paramount importance. The physics of devices and the higher abstraction levels of the digital electronics come together in this context. In [8], M. Favalli et al. discussed the data randomization for reducing or suppressing errors. In this work, they proposed a randomization scheme that is easy to implement, cost effective, and fully scalable with memory dimensions and guarantees optimal randomization along the wordline and the bitline dimensions. The method has been validated on commercial off-the-shelf TLC 3D NAND Flash memory. In [9], Du et al. defined garbage collection (GC) as a time-consuming but necessary operation in Flash memories. They performed a comprehensive experimental study in view of a performance cliff that closely relates to Quality of Service (QoS). Through system-level simulations, they found that 3D NAND Flash based SSDs exacerbate the situation by inducing a much higher number of page migrations during GC. To relieve the performance cliff problem, they propose PreGC to assist normal GC. Experimental results show that PreGC can efficiently relieve the performance cliff by reducing the tail latency from the 90th to 99.99th percentiles.

We thank all the authors who submitted their papers to this Special Issue. We would also like to acknowledge all the reviewers, whose careful and timely reviews ensured the quality of this Special Issue.

Conflicts of Interest: The authors declare no conflict of interest.

References

1. Spinelli, A.; Malavena, G.; Lacaita, A.; Monzio Compagnoni, C. Random Telegraph Noise in 3D NAND Flash Memories. *Micromachines* **2021**, *12*, 703. [CrossRef]
2. Ramesh, S.; Ajaykumar, A.; Ragnarsson, L.; Breuil, L.; El Hajjam, G.; Kaczer, B.; Belmonte, A.; Nyns, L.; Soulié, J.; Van den bosch, G.; et al. Understanding the Origin of Metal Gate Work Function Shift and Its Impact on Erase Performance in 3D NAND Flash Memories. *Micromachines* **2021**, *12*, 1084. [CrossRef]
3. Chen, F.; Chen, B.; Lin, H.; Kong, Y.; Liu, X.; Zhan, X.; Chen, J. Temperature Impacts on Endurance and Read Disturbs in Charge-Trap 3D NAND Flash Memories. *Micromachines* **2021**, *12*, 1152. [CrossRef]
4. Yi, S.; Kim, J. Novel Program Scheme of Vertical NAND Flash Memory for Reduction of Z-Interference. *Micromachines* **2021**, *12*, 584. [CrossRef]
5. Song, Y.; Park, B. Retention Enhancement in Low Power NOR Flash Array with High-κ–Based Charge-Trapping Memory by Utilizing High Permittivity and High Bandgap of Aluminum Oxide. *Micromachines* **2021**, *12*, 328. [CrossRef]
6. Zhang, H.; Wang, J.; Chen, Z.; Pan, Y.; Lu, Z.; Liu, Z. An SVM-Based NAND Flash Endurance Prediction Method. *Micromachines* **2021**, *12*, 746. [CrossRef]
7. He, R.; Hu, H.; Xiong, C.; Han, G. Artificial Neural Network Assisted Error Correction for MLC NAND Flash Memory. *Micromachines* **2021**, *12*, 879. [CrossRef]
8. Favalli, M.; Zambelli, C.; Marelli, A.; Micheloni, R.; Olivo, P. A Scalable Bidimensional Randomization Scheme for TLC 3D NAND Flash Memories. *Micromachines* **2021**, *12*, 759. [CrossRef]
9. Du, Y.; Liu, W.; Gao, Y.; Ausavarungnirun, R. Observation and Optimization on Garbage Collection of Flash Memories: The View in Performance Cliff. *Micromachines* **2021**, *12*, 846. [CrossRef]

 micromachines

Article

Retention Enhancement in Low Power NOR Flash Array with High-κ–Based Charge-Trapping Memory by Utilizing High Permittivity and High Bandgap of Aluminum Oxide

Young Suh Song [1,2] and Byung-Gook Park [1,*]

[1] Department of Electrical and Computer Engineering, Seoul National University, Seoul 08826, Korea; sys1413@snu.ac.kr
[2] Department of Computer Science, Korea Military Academy, Seoul 01805, Korea
* Correspondence: bgpark@snu.ac.kr

Abstract: For improving retention characteristics in the NOR flash array, aluminum oxide (Al_2O_3, alumina) is utilized and incorporated as a tunneling layer. The proposed tunneling layers consist of $SiO_2/Al_2O_3/SiO_2$, which take advantage of higher permittivity and higher bandgap of Al_2O_3 compared to SiO_2 and silicon nitride (Si_3N_4). By adopting the proposed tunneling layers in the NOR flash array, the threshold voltage window after 10 years from programming and erasing (P/E) was improved from 0.57 V to 4.57 V. In order to validate our proposed device structure, it is compared to another stacked-engineered structure with $SiO_2/Si_3N_4/SiO_2$ tunneling layers through technology computer-aided design (TCAD) simulation. In addition, to verify that our proposed structure is suitable for NOR flash array, disturbance issues are also carefully investigated. As a result, it has been demonstrated that the proposed structure can be successfully applied in NOR flash memory with significant retention improvement. Consequently, the possibility of utilizing HfO_2 as a charge-trapping layer in NOR flash application is opened.

Keywords: retention characteristic; high-κ; nonvolatile charge-trapping memory; stack engineering; NOR flash memory; aluminum oxide

Citation: Song, Y.S.; Park, B.-G. Retention Enhancement in Low Power NOR Flash Array with High-κ–Based Charge-Trapping Memory by Utilizing High Permittivity and High Bandgap of Aluminum Oxide. *Micromachines* **2021**, *12*, 328. https://doi.org/10.3390/mi12030328

Academic Editors: Cristian Zambelli and Rino Micheloni

Received: 15 February 2021
Accepted: 15 March 2021
Published: 19 March 2021

1. Introduction

With the advent of the Fifth Generation Mobile Networks (5G) era, the demand for big data has increased rapidly in recent years [1–3], and the need for memory devices enabling more data storage has consistently increased [4,5]. In order to satisfy these demands, novel memory devices utilizing new materials such as aluminum oxide (Al_2O_3, alumina), hafnium oxide (HfO_2), zirconium dioxide (ZrO_2), stacked HfO_2/Al_2O_3, and nano-laminated forms ($HfAlO_x$) have been widely proposed and studied [6–8].

Among them, hafnium oxide (HfO_2) has a tremendous advantage as a charge-trapping layer (CTL) material, since its charge trap density is four times higher than that of the conventional charge-trapping layer (CTL), silicon nitride (Si_3N_4) [9,10]. This enriched CTL density of HfO_2 can enable a wider threshold voltage (V_{TH}) window and improved memory margin [11,12]. Furthermore, permittivity of HfO_2 is much higher than that of Si_3N_4, which enables significant reduction in equivalent oxide thickness (EOT) of the gate stack [13–17]. This enables low program voltage (V_{PGM}), low erase voltage (V_{ERS}), fast program/erase (P/E) speed, fast switching speed, and low power consumption.

From these various advantages of higher charge trap density and the possibility of reducing EOT, HfO_2-based charge-trapping memories (CTM) have been widely studied for fast, high-capacity nonvolatile memory devices [18–21]. However, despite these advantages, HfO_2 has encountered many limitations in commercialization due to retention problems that come from its numerous shallow traps [22–25]. Therefore, this issue needs to be solved for realizing practical high-κ–based charge-trapping memory (HCTM).

In order to solve these retention issues, the use of Al_2O_3 as a CTL in a metal/Al_2O_3/SiO_2/Si (MAOS) structure has been proposed, but it also suffers from retention problems due to vertical leakage current [26,27]. Another previous solution of simply increasing the thickness of tunneling oxide layers has been proposed to mitigate this retention problem; however, this approach concomitantly results in the degradation in P/E speed and subthreshold swing (SS) due to an increase in EOT of the gate stack [28–31]. Furthermore, this approach inevitably increases V_{PGM}, V_{ERS}, and power consumption. Therefore, a new approach is needed to solve these issues.

In this framework, the aim of this paper is to 1) improve retention characteristics of HfO_2-based CTM by using tunneling oxide layers of SiO_2/Al_2O_3/SiO_2 and 2) validate that our proposed structure can be well applied in the NOR flash array, which has been broadly studied for unsupervised learning [32,33]. For validating retention improvement in the proposed memory device structure, it is also compared with the other bandgap engineering (BE) tunneling oxide layers with SiO_2/Si_3N_4/SiO_2 [34–36].

Consequently, it has been demonstrated that the retention characteristics can be significantly improved in a high-κ–based NOR flash memory device by utilizing the advanced tunneling layers with SiO_2/Al_2O_3/SiO_2 on the tunnel field effect transistor (TFET) structure, which has been broadly studied for low power application [37–44]. From an array perspective, it has been demonstrated that the proposed memory device structure is also able to inhibit the programming in unselected cells by bottom gate effect. Namely, we have designed the memory device structure which is free from disturbance issues in the NOR flash array with enhanced retention characteristics.

This paper is organized as follows. First, the basic transfer characteristics are analyzed after calibration. Second, performance of inhibition in the NOR flash array is demonstrated. Then, improvement of the retention characteristics is carefully analyzed with various perspectives. Finally, the expected advantage of applying our proposed memory device structure in the NOR flash array is discussed.

2. Device Structure and Model Physics

2.1. Structure of the Proposed Memory Device

In previous research, the advanced bandgap-engineered TaN/Al_2O_3/HfO_2/SiO_2/Si (BE-TAHOS) structure has been investigated for a faster erasing speed and larger memory window by incorporating Si_3N_4 at the tunneling oxide layer [37–44]. By utilizing this BE-TAHOS structure [34–36] and applying Al_2O_3 at the tunneling layer, the advanced structure of TaN/Al_2O_3/HfO_2/SiO_2/Al_2O_3/SiO_2/Si (TAHOAOS) is designed for NOR flash memory.

Cross-sectional views of conventional TaN/Al_2O_3/HfO_2/SiO_2/Si (TAHOS), BE-TAHOS, and that of the proposed TAHOAOS structure are schematically shown in Figure 1. In order to compare the proposed TAHOAOS structure with not only conventional TAHOS but also the BE-TAHOS structure, BE-TAHOS is also designed with SiO_2/Si_3N_4/SiO_2 tunneling oxide layers [34–36]. The devices designed in this work have four terminals with top gate, bottom gate, source, and drain. The bottom gate is designed for solving disturbance issues.

Table 1 describes the film thickness and channel length for these devices. The simulated devices are composed of tunneling oxide layers with the same EOT of 3 nm for fair comparison. The blocking oxide is composed of 6 nm Al_2O_3, and CTL is composed of 4 nm HfO_2. Bottom gate dielectric has a 3 nm thickness with SiO_2. The length and thickness of the silicon channel are 40 nm and 12 nm, respectively. A gate-drain underlap (gate-source overlap) structure is applied for suppressing ambipolar current [38,39], which undesirably increases the off-state current. In specific, since the ambipolar current occurs due to band-to-band-tunneling (BTBT) current in the body/drain region, it is possible to suppress the ambipolar current by locating the gate far from the drain, which is called gate-drain underlap [38,39].

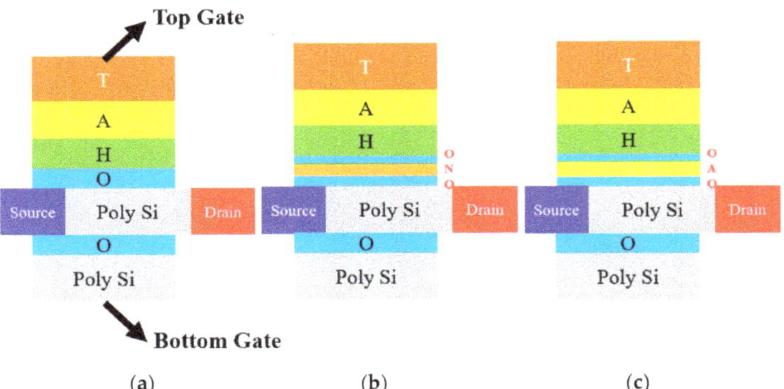

Figure 1. Schematic view illustrating (**a**) conventional $TaN/Al_2O_3/HfO_2/SiO_2/Si$ (TAHOS), (**b**) bandgap engineered (BE)-TAHOS, and (**c**) proposed $TaN/Al_2O_3/HfO_2/SiO_2/Al_2O_3/SiO_2/Si$ (TAHOAOS) structure with two gate terminals. All structures commonly have HfO_2 as charge-trapping layer (CTL) and Al_2O_3 as blocking oxide. The abbreviated letters T, A, H, O, N stand for tantalum nitride (TaN, gate metal), Al_2O_3, HfO_2, SiO_2, Si_3N_4, respectively.

Table 1. Film thickness and channel length in conventional TAHOS, BE-TAHOS, and proposed TAHOAOS structure.

Region	Material	Thickness (nm)
Tunneling oxide	SiO_2	3
	$SiO_2/Si_3N_4/SiO_2$	1/1.7/1
	$SiO_2/Al_2O_3/SiO_2$	1/2.3/1
Blocking oxide	Al_2O_3	6
Charge-trapping layer	HfO_2	4
Bottom gate dielectric	SiO_2	3
Channel (length)	Si	40
Channel (thickness)	Si	12

2.2. Model Physics and Model Parameters

To carefully investigate the electrical characteristics in these three different structures, tunneling models such as band-to-band-tunneling (BTBT), Fowler-Nordheim (FN) tunneling, direct tunneling, and trap-assisted tunneling (TAT) are applied in this device simulation with Synopsys Sentaurus™ through a technology computer-aided design (TCAD) tool. Physical models including Shockley-Read-Hall (SRH) recombination and E-field saturation models are also applied for precisely analyzing the memory operation.

For details, we adopted various mobility models including the PhuMob mobility model, Enormal (Lombardi) mobility model, and thin-layer mobility model to consider interfacial surface calibration roughness scattering and Coulomb scattering. In addition, models of eHighFieldSaturation, hHighFieldSaturation, and Avalanche (CarrierTempDrive) are used for reflecting velocity saturation and avalanche breakdown. Non-local mesh, eBarrierTunneling, and hBarrierTunneling are utilized for applying FN tunneling and direct tunneling.

In modeling HfO_2 as CTL, charge trap density of 1.2×10^{20} cm^{-3} is applied for HfO_2, which corresponds to its charge trap density in memory device [9–11]. Specifically, the energy depth of electron is set as 0.7 eV from the lowest conduction band (LCB) of HfO_2 [20], whereas the energy depth of hole is set as 2.9 eV from the highest valence band (HVB) [21] of HfO_2. On the other hand, in modeling Al_2O_3, charge trap density of 2.0×10^{12} cm^{-3} is applied, and the energy depth of electron/hole is set as 0.4/2.7 eV from LCB/HVB, respectively [8]. In addition, effective electron tunneling masses (m_{eff}) of 0.55 m_o, 0.2 m_o, and 0.4 m_o are used in thin film of SiO_2 [12], HfO_2 [12], and Al_2O_3 [17], respectively.

2.3. Workflow of Study and Calibration Process

Figure 2a illustrates the overall workflow of this paper. The calibration of memory device is performed with the fabricated memory devices [45,46], and then gate dielectric layers of $SiO_2/Al_2O_3/SiO_2$ is incorporated. Thereafter, validation of the proposed memory device structure is conducted in terms of retention characteristics and inhibition in the NOR flash array.

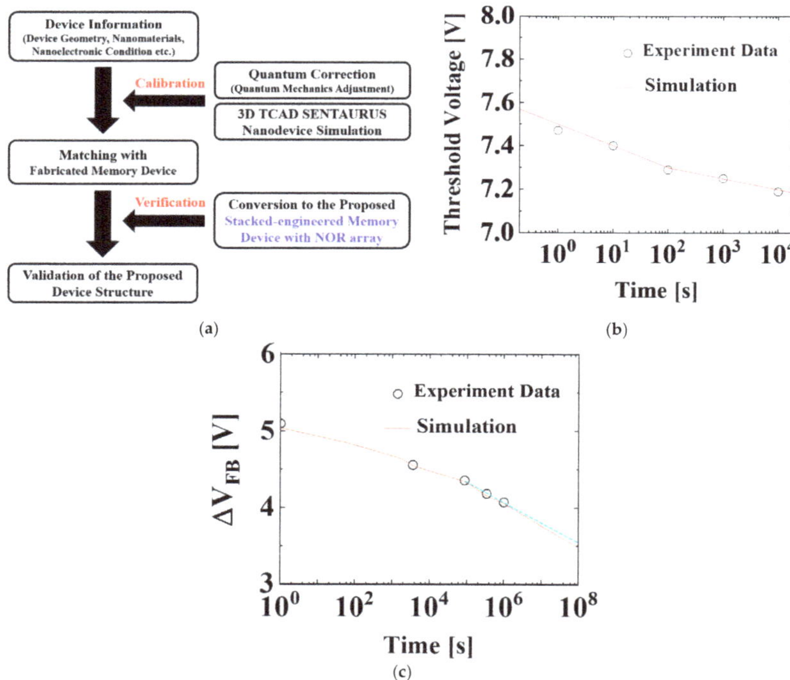

Figure 2. (**a**) Illustration that summarizes overall workflow of this paper; (**b**) calibration results based on the fabricated TANOS device [45]; (**c**) another calibration result based on fabricated BE-TAHOS device [46]. (Sky blue dot line indicates the linear approximation of retention characteristic in the fabricated BE-TAHOS device.).

During the calibration process, quantum correlations are carefully conducted for I_{DS}-I_{GS} calibration, and retention calibration is performed under Synopsys Sentaurus™ three-dimensional (3D) TCAD simulation [47]. For details, we adopted various mobility models including the PhuMob mobility model, Enormal (Lombardi) mobility model, and thin-layer mobility model to consider interfacial surface calibration roughness scattering and Coulomb scattering. Firstly, I_{DS}-I_{GS} calibration is performed by carefully adopting the velocity saturation model, quantum model, and gate work function (WF). Secondly, retention characteristics are carefully calibrated with the fabricated memory devices. Figure 2b,c show our simulation results are well fit with the measured data of retention characteristics in the fabricated $TaN/Al_2O_3/Si_3N_4/SiO_2/Si$ (TANOS) device and BE-TAHOS device.

3. Results and Discussion

3.1. Demonstration of NOR Flash Array with the Proposed Memory Device Structure

Before demonstrating the retention enhancement from the proposed structure, the structure of the proposed memory device must be analyzed. In our proposed device structure, there are two major technological changes.

First, the tunneling oxide layer is technically changed for increasing physical thickness and maintaining the same EOT of 3 nm at the same time (the exact thicknesses are shown in Table 1). Since the EOT of the three structures is the same, the initial transfer characteristics are almost the same, as shown in Figure 3.

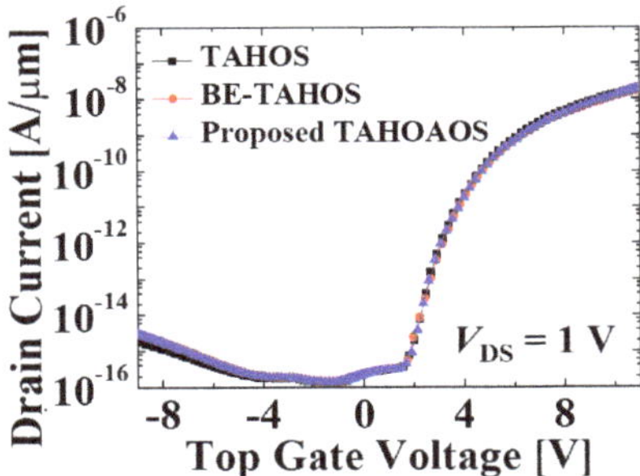

Figure 3. Basic transfer characteristics of three different device structures. These transfer characteristics show that our simulation is well designed with the same EOT thickness.

Second, the bottom gate was added to suppress programming of the unselected cell and solve disturbance issues [37]. Specifically, as illustrated in Figure 4, the additional bottom gates are connected with each other by the bottom gate line, which is perpendicular to the source line and word line. From this perpendicular design between the bottom gate line and word line, it is possible to program the selected cell only and inhibit programming of unselected cells, as described in the following paragraph.

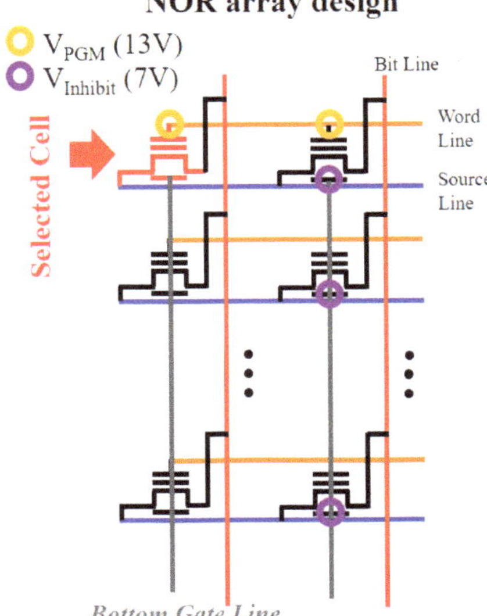

Figure 4. NOR array design for the proposed memory device structure. The newly added bottom gate line is perpendicular to the word line for selective programming.

For programming, the FN tunneling mechanism is used instead of the hot-carrier injection (HCI) mechanism, which has been widely adopted for the conventional programming method in the NOR flash array [48–50]. This is because the conventional HCI programming consumes significant power due to a significant drain current during programming [48]. On the other hand, FN programming can lower power consumption [37] due to its lower gate current compared to the higher drain current during HCI programming [48]. Therefore, FN tunneling is adopted for programming with low power consumption.

Table 2 describes the voltage applied in the selected cell and unselected cells during programming operation under the proposed NOR array design. Programming voltage (V_{PGM}) of 13 V and inhibition voltage of 7 V are adopted, as only 13 V can program the memory cell in high-κ–based memory devices (namely, TAHOS structure) due to low EOT of dielectric layers [18–21].

Table 2. The voltage applied in the selected cell and unselected cells during programming with the proposed NOR array design.

Cell Type	Top Gate Voltage (V)	Bottom Gate Voltage (V)
Selected cell	13	0
Unselected cell 1	13	7
Unselected cell 2	0	7
Unselected cell 3	0	0

The different voltages are applied to the top gate and bottom gate of each cell, which serves as selective programming without disturbance issues. Consequently, as demonstrated in Figure 5, only the selected cell is programmed by FN tunneling, whereas the unselected cells are not. Regarding threshold voltage (Figure 5b), all three unselected cells show nearly zero threshold voltage shift just after programming, whereas the selected sell shows significant threshold voltage shift just after programming. This is because more than 10^{16} cm^{-3} trapped electron charge is needed for threshold voltage shift (Figure 5a,b) [18–21]. Therefore, it is possible to utilize our proposed structure in the NOR flash array without disturbance issues and increase the capacity of memory storage.

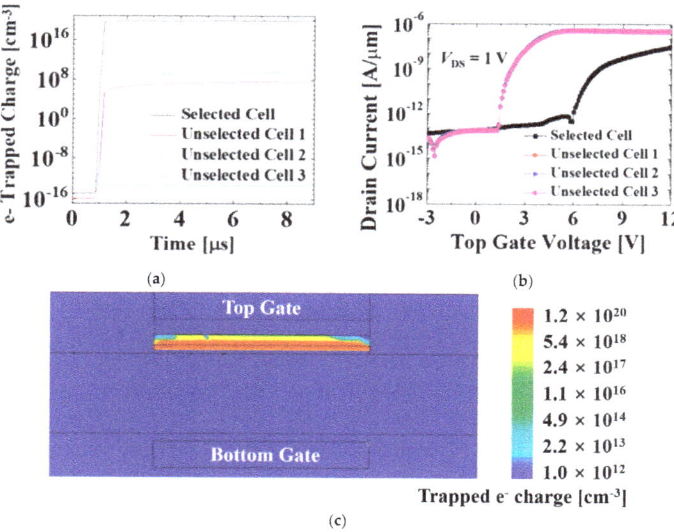

Figure 5. (**a**) Change of electron charge trap density during programming at the cells of the proposed NOR array design. The density of trapped electron charge becomes saturated due to limited top gate voltage. In the selected cell, the higher top gate voltage may increase the saturated density of the trapped electron charge; (**b**) transfer characteristics just after programming of the cells in the proposed NOR array design; (**c**) cross-sectional view of the selected cell with TAHOS structure that illustrates the distribution of the trapped electron charge after programming.

3.2. Retention Enhancement of the Proposed Memory Device Structure

In order to investigate the retention enhancement of the proposed TAHOAOS structure, devices with conventional TAHOS, and BE-TAHOS, proposed TAHOAOS structures are programmed and erased with top gate voltage as described in Figure 6a. Specifically, the high top gate voltage (17 V for programming and −21 V for erasing) is applied in order to perform a fair comparison by matching initial threshold voltage (namely, threshold voltage when time is 10^{-3} s). Then, retention characteristics of each structure are analyzed for 10 years. It is shown that our proposed TAHOAOS structure maintains a significant threshold voltage window for 10 years and is very strategic for retention characteristics, as demonstrated in Figure 6b.

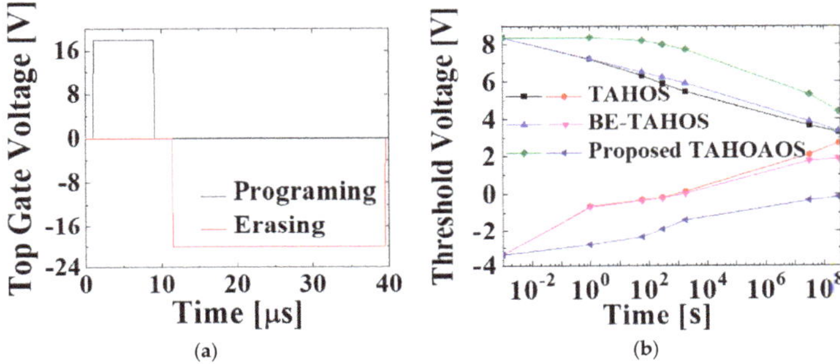

Figure 6. (**a**) Top gate bias during programming and erasing, and (**b**) retention characteristics of the conventional TAHOS, BE-TAHOS, and the proposed TAHOAOS structure. The high top gate voltage (17 V for programming and −21 V for erasing) is applied in order to perform fair comparison by matching initial threshold voltage at 1 micro-second. (Specifically, programming with top gate voltage of 13 V, as in Table 2, results in different initial threshold voltage [37], and hence programming with a higher top gate voltage of 17 V is performed for fair comparison.).

Specifically, our proposed TAHOAOS structure maintains 4.57 V of the threshold voltage window, whereas conventional TAHOS structure only maintains 0.57 V after 10 years from programming and erasing (P/E) as illustrated in Figure 7. It is remarkable that our proposed TAHOAOS structure shows better retention characteristics (more than three times) compared to the BE-TAHOS structure.

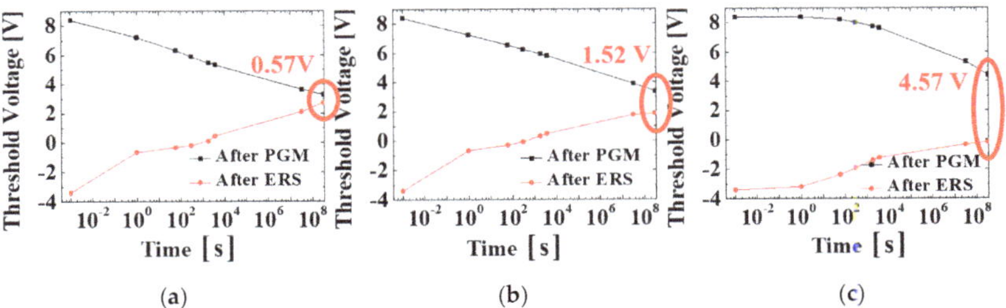

Figure 7. Detailed description of retention characteristics in (**a**) conventional TAHOS, (**b**) BE-TAHOS, and (**c**) proposed TAHOAOS structure.

However, there is one remarkable point in these retention characteristics. As shown in Figure 6b, the retention characteristics of conventional TAHOS and BE-TAHOS after erase operation (namely, red and pink line in Figure 6b) show barely little difference. Namely, even though retention characteristics of BE-TAHOS (pink line) is slightly better than that of conventional TAHOS (red line), the difference between them is reduced due to valence band offset.

This can be explained by energy band diagram. Figure 8 shows the energy band diagram of BE-TAHOS and the proposed TAHOAOS structure with reference to previous fabricated devices of the TAHOS and TANOS structure [51]. As illustrated in Figure 8a, substantial valence band offset exists in the BE-TAHOS structure. This valence band offset helps the hole to be ejected from HfO_2 CTL. Therefore, the advantage of thicker tunneling oxide layers in BE-TAHOS (compared to conventional TAHOS) is reduced in terms of retention characteristics.

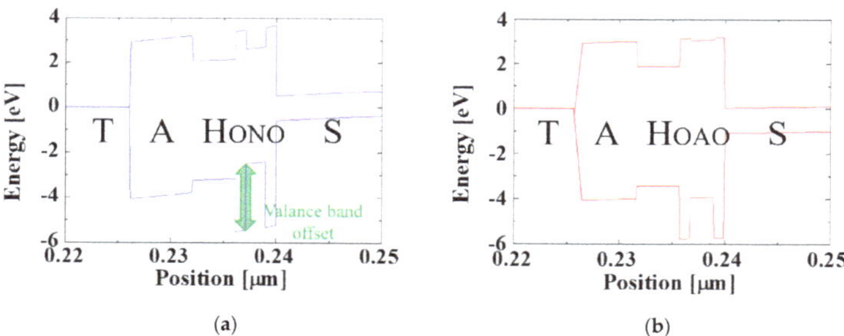

Figure 8. Energy band diagram of (**a**) BE-TAHOS, and (**b**) proposed TAHOAOS structure. Regarding retention characteristics, the valence band offset of BE-TAHOS (green arrow in panel a) mitigates the advantage of thicker tunneling oxide layers in BE-TAHOS. The abbreviated letters T, A, H, O, N stand for tantalum nitride (TaN, gate metal), Al_2O_3, HfO_2, SiO_2, Si_3N_4, respectively.

On the other hand, the proposed TAHOAOS structure has not only thicker tunneling oxide layers but also lower valance band offset compared to BE-TAHOS (Figure 8). Therefore, regarding hole retention, the proposed TAHOAOS structure has a remarkable competitive edge, as demonstrated in Figure 6b.

Figure 9 shows the transfer curves after 10 years of P/E operation in the conventional TAHOS structure and the proposed TAHOAOS structure. It is expected that our proposed structure can serve as a powerful tool for future big data markets with better reliability (retention), higher memory capacity, and low power operation (TFET-based memory [34–40]).

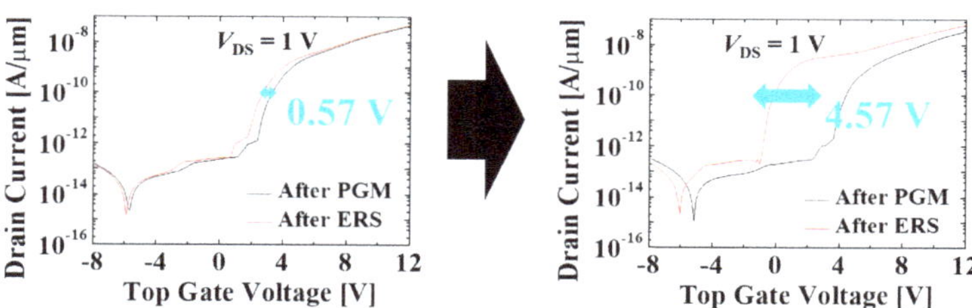

Figure 9. Enhancement of retention characteristics by the proposed tunneling oxide engineering. The graphs are calculated after 10 years of programming and erasing.

In summary, we have improved the retention characteristics with which HfO_2-based nonvolatile charge-trapping memory has encountered [22–25], and opened up the possibility of practical application of HfO_2-based NOR flash memory for better memory capacity.

3.3. Proposal for Future Research

We have proposed the design methodology for better retention characteristics and great immunity against disturbance issues by developing the TAHOAOS structure [37] on the NOR flash array. The proposed design technology is expected to improve the retention characteristics and decrease power consumption during programming (due to the programming method of FN tunneling) and during read operation (due to the TFET-based structure). Furthermore, it is expected that our newly proposed device structure with four terminals can solve the disturbance issue and make only a selected cell programmed.

However, even though our research has made considerable efforts to verify our proposed methodology, our research is basically limited to NOR flash application. We believe our proposed TAHOAOS structure can be applied beyond NOR flash application and to other fields such as 3D NAND flash and 3D AND flash. This is because our proposed technology may be applied in another domain by changing the design of the circuit. Therefore, we would like to suggest the future research topic to readers by analyzing our proposed technique in another array and another circuit design. It may be a desirable and interesting topic to develop our research with various future memory applications.

4. Conclusions

In this study, we propose the advanced structure for the NOR flash array with retention improvement. From the bottom gate effect, the disturbance issues are well suppressed, and it is possible to utilize the proposed structure in a NOR flash array. In addition, the threshold voltage window after 10 years of programming and erasing was considerably increased from 0.57 V to 4.57 V by incorporating Al_2O_3 in tunneling oxide layers. This enhancement is achieved by 1) high physical thickness of tunneling layers in the proposed structure (namely, high permittivity of Al_2O_3) and 2) lower valence band offset/conduction band offset in the proposed structure (namely, higher bandgap of Al_2O_3 compared to Si_3N_4). These results open up the possibility of using enriched CTL (HfO_2) with improved retention characteristics. Therefore, the proposed TAHOAOS structure is very strategic for future highly integrated memory cells in big data markets with significant reliability enhancement.

Author Contributions: Writing—Original Draft & Data curation, Y.S.S.; Writing—Review & Editing, B.-G.P.; Validation, Y.S.S.; Supervision B.-G.P. All authors have read and agreed to the published version of the manuscript.

Funding: This work was supported by 2021 research fund of Korea Military Academy (Hwarangdae Research Institute).

Conflicts of Interest: The authors declare no conflict of interest.

References

1. Imran, A.; Zoha, A.; Abu-Dayya, A. Challenges in 5G: How to empower SON with big data for enabling 5G. *IEEE Netw.* **2014**, *28*, 27–33. [CrossRef]
2. Zheng, K.; Yang, Z.; Zhang, K.; Chatzimisios, P.; Yang, K.; Xiang, W. Big data-driven optimization for mobile networks toward 5G. *IEEE Netw.* **2016**, *30*, 44–51. [CrossRef]
3. Han, Q.; Liang, S.; Zhang, H. Mobile cloud sensing, big data, and 5G networks make an intelligent and smart world. *IEEE Netw.* **2015**, *29*, 40–45. [CrossRef]
4. Zhang, H.; Chen, G.; Ooi, B.C.; Tan, K.; Zhang, M. In-Memory Big Data Management and Processing: A Survey. *IEEE Trans. Knowl. Data Eng.* **2015**, *27*, 1920–1948. [CrossRef]
5. Wang, Y.; Yu, H. An ultralow-power memory-based big-data computing platform by nonvolatile domain-wall nanowire devices. In Proceedings of the International Symposium on Low Power Electronics and Design (ISLPED), Beijing, China, 4–6 September 2013.
6. Hamdioui, S.; Xie, L.; Anh Du Nguyen, H.; Taouil, M.; Bertles, K.; Corporaal, H.; Jiao, H.; Cathoor, F.; Wouters, D.; Eike, L.; et al. Memristor based computation-in-memory architecture for data-intensive applications. In Proceedings of the Design, Automation & Test in Europe Conference & Exhibition (DATE), Grenoble, France, 9–13 March 2015.
7. Yoon, S.-M.; Lee, N.-Y.; Ryu, S.-O.; Choi, K.-J.; Park, Y.-S.; Lee, S.-Y.; Yu, B.-G.; Kang, M.-J.; Choi, S.-J.; Wuttig, M. Sb-Se-based phase-change memory device with lower power and higher speed operations. *IEEE Electron. Device Lett.* **2006**, *27*, 445–447. [CrossRef]
8. Marcon, D.; Van Hove, M.; De Jaeger, B.; Posthuma, N.; Wellekens, D.; You, S.; Kang, X.; Wu, T.-L.; Willems, M.; Stoffels, S.; et al. Direct comparison of GaN-based e-mode architectures (recessed MISHEMT and p-GaN HEMTs) processed on 200 mm GaN-on-Si with Au-free technology. In Proceedings of the SPIE 9363, Gallium Nitride Materials and Devices X, San Francisco, CA, USA, 13 March 2015.

9. Zhu, W.J.; Ma, T.P.; Zafar, S.; Tamagawa, T. Charge trapping in ultrathin hafnium oxide. *IEEE Electron. Device Lett.* **2002**, *23*, 597–599. [CrossRef]
10. Zou, X.; Jin, L.; Yan, L.; Zhang, Y.; Ai, D.; Zhao, C.; Xu, F.; Li, C.; Huo, Z. The influence of grain boundary interface traps on electrical characteristics of top select gate transistor in 3D NAND flash memory. *Solid State Electron.* **2019**, *153*, 67–73. [CrossRef]
11. Yao, Y.; Li, C.; Huo, Z.L.; Liu, M.; Zhu, C.X.; Gu, C.Z.; Duan, X.F.; Wang, Y.G.; Gu, L.; Yu, R.C. In situ electron holography study of charge distribution in high-κ charge-trapping memory. *Nat. Commun.* **2013**, *4*, 1–8. [CrossRef]
12. Hinkle, C.L.; Fulton, C.; Nemanich, R.J.; Lucovsky, G. A novel approach for determining the effective tunneling mass of electrons in HfO2 and other high-K alternative gate dielectrics for advanced CMOS devices. *Microelectron. Eng.* **2004**, *72*, 257–262. [CrossRef]
13. Ali, T.; Polakowski, P.; Riedel, S.; Büttner, T.; Kämpfe, T.; Rudolph, M.; Pätzold, B.; Seidel, K.; Löhr, D.; Hoffmann, R.; et al. High Endurance Ferroelectric Hafnium Oxide-Based FeFET Memory Without Retention Penalty. *IEEE Trans. Electron. Devices* **2018**, *65*, 3769–3774. [CrossRef]
14. Lee, B.H.; Kang, L.; Qi, W.J.; Nieh, R.; Jeon, Y.; Onishi, K.; Lee, J.C. Ultrathin hafnium oxide with low leakage and excellent reliability for alternative gate dielectric application. In Proceedings of the International Electron Devices Meeting (IEDM), Washington, DC, USA, 5–8 December 1999.
15. Choi, Y.; Lee, K.; Kim, K.Y.; Kim, S.; Lee, J.; Lee, R.; Kim, H.-M.; Song, Y.S.; Kim, S.; Lee, J.-H.; et al. Simulation of the effect of parasitic channel height on characteristics of stacked gate-all-around nanosheet FET. *Solid State Electron.* **2020**, *164*, 107686. [CrossRef]
16. Kim, J.H.; Kim, H.W.; Song, Y.S.; Kim, S.; Kim, G. Analysis of Current Variation with Work Function Variation in L-Shaped Tunnel-Field Effect Transistor. *Micromachines* **2020**, *11*, 780. [CrossRef] [PubMed]
17. Molina-Reyes, J.; Uribe-Vargas, H.; Torres-Torres, R.; Mani-Gonzalez, P.G.; Herrera-Gomez, A. Accurate modeling of gate tunneling currents in Metal-Insulator-Semiconductor capacitors based on ultra-thin atomic-layer deposited Al_2O_3 and post-metallization annealing. *Thin Solid Film.* **2017**, *638*, 48–56. [CrossRef]
18. Xu, W.C.; He, H.X.; Jing, X.S.; Wu, S.J.; Zhang, Z.; Gao, J.W.; Gao, X.S.; Zhou, G.F.; Lu, X.B.; Liu, J.-M. High performance organic nonvolatile memory transistors based on HfO_2 and poly(α-methylstyrene) electret hybrid charge-trapping layers. *Appl. Phys. Lett.* **2017**, *111*, 1–5. [CrossRef]
19. You, H.-C.; Kuo, P.-Y.; Ko, F.-H.; Chao, T.-S.; Lei, T.-F. The Impact of Deep Ni Salicidation and hbox NH_3 Plasma Treatment on Nano-SOI FinFETs. *IEEE Electron. Device Lett.* **2006**, *27*, 799–801. [CrossRef]
20. Driussi, F.; Spiga, S.; Lamperti, A.; Congedo, G.; Gambi, A. Simulation Study of the Trapping Properties of HfO_2-Based Charge-Trap Memory Cells. *IEEE Trans. Electron. Devices* **2014**, *61*, 2056–2063. [CrossRef]
21. Zhang, Y.; Shao, Y.Y.; Lu, X.B.; Zeng, M.; Zhang, Z.; Gao, X.S.; Zhang, X.J.; Liu, J.-M.; Dai, J.Y. Defect states and charge trapping characteristics of HfO_2 films for high performance nonvolatile memory applications. *Appl. Phys. Lett.* **2014**, *105*, 172902. [CrossRef]
22. Stesmans, A.; Afanasev, V.V. Defect correlated with positive charge trapping in functional HfO_2 layers on (100)Si revealed by electron spin resonance: Evidence for oxygen vacancy? *Microelectron. Eng.* **2017**, *178*, 112–115. [CrossRef]
23. Chen, Y.Y.; Goux, L.; Clima, S.; Govoreanu, B.; Degraeve, R.; Sankar Kar, G.; Fantini, A.; Groesenken, G.; Wouters, D.J.; Jurczak, M.; et al. Endurance/Retention Trade-off on HfO2/Metal Cap 1T1R Bipolar RRAM. *IEEE Trans. Electron. Devices* **2013**, *60*, 1114–1121. [CrossRef]
24. Mitrovic, I.Z.; Luy, Y.; Buiu, O.; Hall, S. Current transport mechanisms in HfO_2xSiO_21-x/SiO_2gate stacks. *Microelectron. Eng.* **2007**, *84*, 2306–2309. [CrossRef]
25. Cerbu, F.; Madia, O.; Andreev, D.V.; Fadida, S.; Eizenberg, M.; Breuil, L.; Lisoni, J.G.; Kittl, J.A.; Strand, J.; Shluger, A.L.; et al. Intrinsic electron traps in atomic-layer deposited HfO2 insulators. *Appl. Phys. Lett.* **2016**, *108*, 1–5. [CrossRef]
26. Bersuker, G.; Sim, J.H.; Park, C.S.; Young, C.D.; Nadkarni, S.V.; Choi, R.; Lee, H.B. Mechanism of Electron Trapping and Characteristics of Traps in HfO2 Gate Stacks. *IEEE Trans. Device Mater. Reliab.* **2007**, *7*, 138–145. [CrossRef]
27. Khosla, R.; Rolseth, E.G.; Kumar, P.; Vadakupudhupalayam, S.S.; Sharma, S.K.; Schulze, J. Charge Trapping Analysis of Metal/Al2O3/SiO2/Si, Gate Stack for Emerging Embedded Memories. *IEEE Trans. Device Mater. Reliab.* **2017**, *17*, 80–89. [CrossRef]
28. Xu, Z.; Huo, Z.; Zhu, C.; Cui, Y.; Wang, M.; Zheng, Z.; Liu, J.; Wang, Y.; Li, F.; Liu, M. Performance-improved nonvolatile memory with aluminum nanocrystals embedded in Al2O3 for high temperature applications. *J. Appl. Phys.* **2011**, *110*, 1–5. [CrossRef]
29. Gu, S.-H.; Hsu, C.-W.; Wang, T.; Lu, W.-P.; Ku, Y.-H.J.; Lu, C.-Y. Numerical Simulation of Bottom Oxide Thickness Effect on Charge Retention in SONOS Flash Memory Cells. *IEEE Trans. Electron. Devices* **2006**, *54*, 90–97. [CrossRef]
30. King, Y.-C.; Liu, T.-J.K.; Hu, C. A long-refresh dynamic/quasi-nonvolatile memory device with 2-nm tunneling oxide. *IEEE Electron. Device Lett.* **1999**, *20*, 409–411. [CrossRef]
31. Hanafi, H.; Tiwari, S.; Khan, I. Fast and long retention-time nano-crystal memory. *IEEE Trans. Electron. Devices* **1996**, *43*, 1553–1558. [CrossRef]
32. Malavena, G.; Spinelli, A.S.; Compagnoni, C.M. Implementing Spike-Timing-Dependent Plasticity and Unsupervised Learning in a Mainstream NOR Flash Memory Array. In Proceedings of the 2018 IEEE International Electron Devices Meeting (IEDM), San Francisco, CA, USA, 1–5 December 2018; pp. 1–4.

33. Malavena, G.; Filippi, M.; Spinelli, A.S.; Compagnoni, C.M. Unsupervised Learning by Spike-Timing-Dependent Plasticity in a Mainstream NOR Flash Memory Array—Part I: Cell Operation. *IEEE Trans. Electron. Devices* **2019**, *66*, 4727–4732. [CrossRef]
34. Lue, H.-T.; Wang, S.Z.; Lai, E.-K.; Shih, Y.-H.; Lai, S.-C.; Yang, L.-W.; Chen, K.-C.; Ku, J.; Hsieh, K.-Y.; Liu, R.; et al. BE-SONOS: A bandgap engineered SONOS with excellent performance and reliability. In Proceedings of the IEEE International Electron Devices Meeting, Washington, DC, USA, 5 December 2005.
35. Lue, H.-T.; Wang, S.-Y.; Lai, E.-K.; Hsieh, K.-Y.; Liu, R.; Lu, C.Y. A BE-SONOS (Bandgap Engineered SONOS) NAND for Post-Floating Gate Era Flash Memory. In Proceedings of the 2007 International Symposium on VLSI Technology, Systems and Applications (VLSI-TSA), Hsinchu, Taiwan, 23–25 April 2007; pp. 1–2.
36. Hsu, T.-H.; Lue, H.T.; King, Y.-C.; Hsieh, J.-Y.; Lai, E.-K.; Hsieh, K.-Y.; Liu, R.; Lu, C.-Y. A High-Performance Body-Tied FinFET Bandgap Engineered SONOS (BE-SONOS) for nand-Type Flash Memory. *IEEE Electron. Device Lett.* **2007**, *28*, 443–445. [CrossRef]
37. Song, Y.S.; Jang, T.; Min, K.K.; Baek, M.-H.; Yu, J.; Kim, Y.; Lee, J.-H.; Park, B.-G. Tunneling oxide engineering for improving retention in nonvolatile charge-trapping memory with TaN/Al_2O_3/HfO_2/SiO_2/Al_2O_3/SiO_2/Si structure. *Jpn. J. Appl. Phys.* **2020**, *59*, 61006. [CrossRef]
38. Khatami, Y.; Banerjee, K. Steep Subthreshold Slope n- and p-Type Tunnel-FET Devices for Low-Power and Energy-Efficient Digital Circuits. *IEEE Trans. Electron. Devices* **2009**, *56*, 2752–2761. [CrossRef]
39. Yu, J.; Kim, S.; Baek, M.-H.; Min, K.K.; Jang, T.; Song, Y.S.; Park, B.G. Investigation of Ambipolar Current Suppression Using Dual Work Function Metal Gate in L-Shaped Tunnel Field Effect Transistor. In Proceedings of the IEIE Summer Conference, Jeju-do, Korea, 26–28 June 2019.
40. Kim, Y.; Kim, T.; Beak, M.H.; Jang, T.; Song, Y.S.; Jeon, B.; Park, B.G. An Area Efficient Adaptive Neuron Circuit Exploiting Tunnel Field-Effect Transistor. In Proceedings of the IEIE Summer Conference, Jeju-do, Korea, 26–28 June 2019.
41. Tripathy, M.R.; Singh, A.K.; Samad, A.; Chander, S.; Baral, K.; Singh, P.K.; Jit, S. Device and Circuit-Level Assessment of GaSb/Si Heterojunction Vertical Tunnel-FET for Low-Power Applications. *IEEE Trans. Electron. Devices* **2020**, *67*, 1285–1292. [CrossRef]
42. Kim, T.; Park, K.; Jang, T.; Baek, M.-H.; Song, Y.S.; Park, B.-G. Input-modulating adaptive neuron circuit employing asymmetric floating-gate MOSFET with two independent control gates. *Solid State Electron.* **2020**, *163*, 107667. [CrossRef]
43. Kim, T.; Song, Y.S.; Park, B.-G. Overflow Handling Integrate-and-Fire Silicon-on-Insulator Neuron Circuit Incorporating a Schmitt Trigger Implemented by Back-Gate Effect. *J. Nanosci. Nanotechnol.* **2019**, *19*, 6183–6186. [CrossRef]
44. Song, Y.S.; Kim, J.H.; Kim, G.; Kim, H.-M.; Kim, S.; Park, B.-G. Improvement in Self-Heating Characteristic by Incorporating Hetero-Gate-Dielectric in Gate-All-Around MOSFETs. *IEEE J. Electron. Devices Soc.* **2021**, *9*, 36–41. [CrossRef]
45. Park, H.; Bersuker, G.; Gilmer, D.; Lim, K.Y.; Jo, M.; Hwang, H.; Padovani, A.; Larcher, L.; Pavan, P.; Taylor, W.; et al. Charge loss in TANOS devices caused by Vt sensing measurements during retention. In Proceedings of the 2010 IEEE International Memory Workshop, Seoul, Korea, 16–19 May 2010; pp. 1–2. [CrossRef]
46. Congedo, G.; Lamperti, A.; Salicio, O.; Spiga, S. Multi-Layered Al_2O_3/HfO_2/SiO_2/Si_3N_4/SiO_2 Thin Dielectrics for Charge Trap Memory Applications. *ECS J. Solid State Sci. Technol.* **2013**, *2*, N1–N5. [CrossRef]
47. Synopsys Inc. *Version K-2015.06-SP*; Synopsys Inc.: Mountain View, CA, USA, 2017.
48. Della Marca, V.; Postel-Pellerin, J.; Just, G.; Canet, P.; Ogier, J.-L. Impact of endurance degradation on the programming efficiency and the energy consumption of NOR flash memories. *Microelectron. Reliab.* **2014**, *54*, 2262–2265. [CrossRef]
49. Compagnoni, C.M.; Chiavarone, L.; Calabrese, M.; Ghidotti, M.; Lacaita, A.L.; Spinelli, A.S.; Visconti, A. Fundamental Limitations to the Width of the Programmed VTVT Distribution of nor Flash Memories. *IEEE Trans. Electron. Devices* **2010**, *57*, 1761–1767. [CrossRef]
50. Choi, S.-J.; Han, J.-W.; Kim, S.; Moon, D.-I.; Jang, M.-G.; Kim, J.S.; Kim, K.H.; Lee, G.S.; Oh, J.S.; Song, M.H.; et al. Performance Breakthrough in NOR Flash Memory with Dopant-Segregated Schottky-Barrier (DSSB) SONOS Devices. In Proceedings of the 2009 Symposium on VLSI Technology, Kyoto, Japan, 15–17 June 2009.
51. Spiga, S.; Congedo, G.; Russo, U.; Lamperti, A.; Salicio, O.; Driussi, F.; Vianello, E. Experimental and simulation study of the program efficiency of HfO2 based charge trapping memories. In Proceedings of the 2010 European Solid State Device Research Conference, Sevila, Spain, 14–16 September 2010; pp. 408–411.

 micromachines

Article

Novel Program Scheme of Vertical NAND Flash Memory for Reduction of Z-Interference

Su-in Yi [1,2] and Jungsik Kim [3,4,*]

[1] Department of Electrical and Computer Engineering, Texas A&M University, College Station, TX 77843, USA; yisuin@tamu.edu
[2] Samsung Electronics, Hwasung 18448, Kyeonggi, Korea
[3] Department of Electrical Engineering, Gyeongsang National University, Jinju 52828, Gyeongnam, Korea
[4] Engineering Research Institute (ERI), Gyeongsang National University, Jinju 52828, Gyeongnam, Korea
* Correspondence: jungsik@gnu.ac.kr; Tel.: +82-55-772-1718

Abstract: Minimizing the variation in threshold voltage (V_t) of programmed cells is required to the extreme level for realizing multi-level-cells; as many as even 5 bits per cell recently. In this work, a recent program scheme to write the cells from the top, for instance the 170th layer, to the bottom, the 1st layer, (T-B scheme) in vertical NAND (VNAND) Flash Memory, is investigated to minimize V_t variation by reducing Z-interference. With the aid of Technology Computer Aided Design (TCAD) the Z-Interference for T-B (84 mV) is found to be better than B-T (105 mV). Moreover, under scaled cell dimensions (e.g., L_g: 31→24 nm), the improvement becomes protruding (T-B: 126 mV and B-T: 162 mV), emphasizing the significance of the T-B program scheme for the next generation VNAND products with the higher bit density.

Keywords: NAND flash memory; interference; Technology Computer Aided Design (TCAD) simulation; disturbance; program; non-volatile memory (NVM)

Citation: Yi, S.-i.; Kim, J. Novel Program Scheme of Vertical NAND Flash Memory for Reduction of Z-Interference. *Micromachines* **2021**, *12*, 584. https://doi.org/10.3390/mi12050584

Academic Editors: Cristian Zambelli and Rino Micheloni

Received: 15 April 2021
Accepted: 16 May 2021
Published: 20 May 2021

1. Introduction

Due to the nature of NAND flash memory, which lacks the capability of random access [1] of NOR flash memory [2,3] or other memories such as DRAM (Dynamic Random Access Memory) and PCM (Phase Change Memory), reading and writing operations of one cell inevitably accompanies operations on the other cells simultaneously in a target NAND string [4,5]. Various combinations of the operation scheme such as bit line voltage (V_{BL}), read voltage (V_{READ}), pass voltage (V_{PASS}), etc., are typically tested and finally the optimal set is chosen by product engineers to minimize the threshold voltage (V_t) variation for the given as-fab-out chips [6–9]. Moreover, with the higher level of layers emerging every year or two, such that Memory companies announced a 6th generation vertical NAND (VNAND) flash memory product of 120 layers in 2019 and subsequently plan to announce the next 7th generation of 170 or more layers in a year or so [10], even more complicated combinations of the operation scheme are being developed. For example, varying bias conditions depending on the word line (WL) number, due to the nature of high aspect ratio contact etching [11–13], need to be investigated by trial and error to meet the criteria of V_t variation in a tight schedule. For this reason, the operation scheme optimization process heavily relies on the product engineers' intuition or, recently, statistical approaches such as machine learning technology which can often neglect to understand the underlying charge transport physics [14,15]. However, in order to accumulate the prior experience on the operation scheme optimization toward the sustainable technique for future products, it is critically important to understand the correlation between the input (operation scheme) and the output (V_t variation).

2. Simulation Methods

In this report, we target the investigation of V_t interference and coupling dependency on the programming direction in a bit line as shown in Figure 1a. One method is to program beginning from the bottom to top (B-T), i.e., from WL1 to WL170, which is the scheme adopted by early generations of VNAND, and the other is to program beginning from the top to bottom (T-B), i.e., from WL170 to WL1, which has been recently employed [16–21]. Although the scheme of T-B is currently prevailing over B-T because of the better vulnerability toward interference/coupling, as mentioned earlier, this link may have been found through empirical trials based on a few prior reports with outdated cell geometries [22,23]. That is probably the reason why any quantitative analysis and investigation is unavailable publicly with up-to-date VNAND cell structures. In this work, by performing Technology-Computer Aided Design (TCAD) simulations (Synopsys^{TM}, Mountain View, CA, USA) of interference for the two distinct schemes [24], we provide solid understanding on the difference between the two and evaluate the benefit for the scaled-down cells of next generation VNAND products.

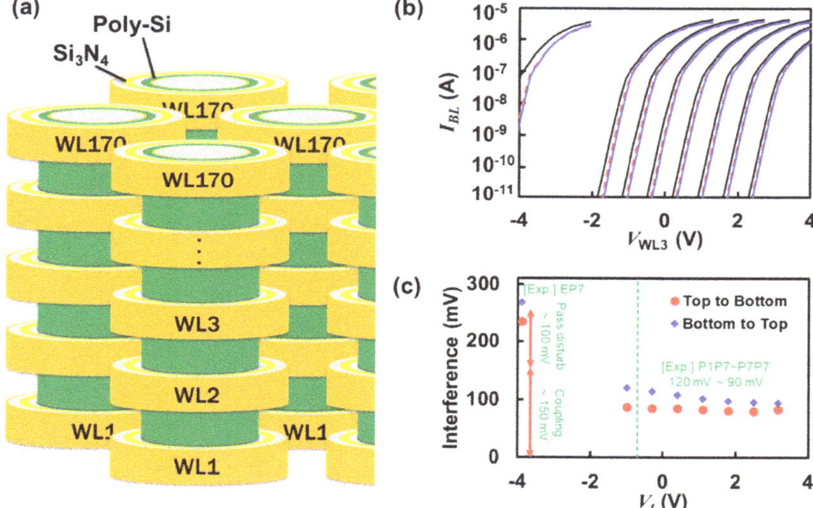

Figure 1. (**a**) Vertical NAND (VNAND) Flash cell array schematic showing neighbors both intra-string (Z-interference) and inter-string. Z-interference is the most critical since the channel is shared in close proximity for VNAND products. (**b**) Current versus voltage data as a function of the gate voltage of the victim cell (WL3). Solid line represents the reference state before the aggressor cell is programmed. Red dashed line and blue solid line denote the states after interference by T-B and B-T, respectively. (**c**) Interference of 8 different states (E, P1, P2, P3, P4, P5, P6, and P7) for triple level cell (TLC) under the condition of the aggressor programmed to P7 (V_t = 3.177 V). Blue diamonds and red circles show the results out of Bottom to Top (B-T, WL4 is aggressor) and Top to Bottom (T-B, WL2 is aggressor), respectively. Remarks with (Exp.) denote experimentally measured interferences (unpublished) from Samsung's 4th generation VNAND (Ref. [18]).

3. Results and Discussion

Figure 1b shows eight sets of I_{BL}-V_{WL} curves at V_t's from the erased state (E) to the programmed states (P1, P2, ... , P6, and P7). The tunneling masses of 0.36 m_0 and 0.38 m_0 were used for electrons and holes, and the block erasing with V_{ERS} = −16 V for 1 ms resulted in $V_{t,E}$ = −3.889 V based on BL current I_{BL} = 100 nA. It should be noted that the electron tunneling mass of 0.36 m_0 was chosen to properly describe the pass disturb under EP7 interference (approximately 100 mV of pass disturb and 150 mV of coupling), while this brings about the programming speed faster (Vpgm = 16 V for 100 μs makes $V_{t,P7}$)

compared to experimental results (Vpgm = 19~20 V for 100 µs makes $V_{t,P7}$). This is a well-known dilemma for Flash memory TCAD simulations, where the trap-assisted-tunneling (TAT) model is rarely considered due to the complexity in describing the atom defects in the actually fabricated ONO (Oxide-Nitride-Oxide) films. Moreover, uncertainties due to random telegraph noise (RTN) were not considered to clarify the comparison by mean V_t's [25]. Read voltage (V_{READ}) of 7 V was used as default. Once every cell in the model (five word lines) was written to the state E by the block erasing, seven different programmed states were mimicked by using the programming voltage (V_{PGM}) of 16.0 V, 15.3 V, 14.6 V, 13.9 V, 13.2 V, 12.5 V, and 11.8 V for P7, P6, P5, P4, P3, P2, and P1 states, respectively, on the third word line (WL3) together with V_{PASS} applied to the other cells of 8 V for 100 µs. Consequently, the V_t's of seven programmed states constituted 3.177 V, 2.487 V, 1.794 V, 1.098 V, 0.399 V, −0.293 V, −0.990 V, of which the average read window between two adjacent states is approximately 0.7 V, enabling the triple level cell (TLC). The interference on WL3 was simulated under two different scenarios. The first is when the upper adjacent cell, WL4, is programmed to P7, named as B-T scheme and represented by blue solid lines. The second is when the aggressor is WL2, named as T-B scheme and represented by red dashed lines.

Based on the raw data available in Figure 1b, the amount of interference in the unit of mV as a function programmed state in V_t is rearranged in Figure 1c, which can be labelled as EP7, P1P7, P2P7, P3P7, P4P7, P5P7, P6P7, and P7P7. The green dashed line at V_t = −0.690 V is of the virgin state and it should be noted that the fixed charge of -10^{12} cm^{-2} was used between the poly-silicon channel and fill oxide (the core oxide of a NAND string due to macaroni-like structure) to fit the typical virgin V_t ranging from −0.5 V to +0.5 V. The comparison between T-B (red circles) and B-T (blue diamonds) clearly provides the better interference performance of T-B over B-T. The interference in NAND Flash consists of two contributions: one is the change in trapped charge concentration of the victim cell due to V_{PASS} = 8 V during the programming phase (pass disturb), and the other is the influence of the adjacent cell during the reading phase (coupling). In addition, the distinctively high interference of EP7, 269 mV for B-T and 235 mV for T-B, compared to those of P1P7~P7P7 implies the significant contribution of pass disturb.

Figure 2 provides the net charge concentration (Q_{CON}) information with color plots (a) and curves (b) as a function of the position in the radial axis. Note that r = 0 nm corresponds to the center of the cylindrical symmetry for a VNAND string. Because the diameter of the hole was used to be 120 nm followed by 7.5 nm blocking oxide, 6 nm trap-nitride, and 5.5 nm tunneling oxide, r = 46.5 nm and r = 52.5 nm represent the interfaces with tunneling oxide and blocking oxide, respectively. In this work, we did not consider the tendency of decreasing hole-diameters and ONO film thicknesses with decreasing WL numbers in so called "stack-coverage". Although it is known to cause the variation in threshold voltages of 3D NAND cells [26,27], recent advances in high-aspect-ratio thin film technique produce very decent stack-coverages (ONO film > 95% and Poly-Si > 90% by comparing the film thickness of WL1 to that of WL170). Moreover, the state-of-the-art high-aspect-ratio-etching technique makes almost uniform hole diameters (~120 nm) except for approximately 10% of the top and bottom layers of a NAND string [21].

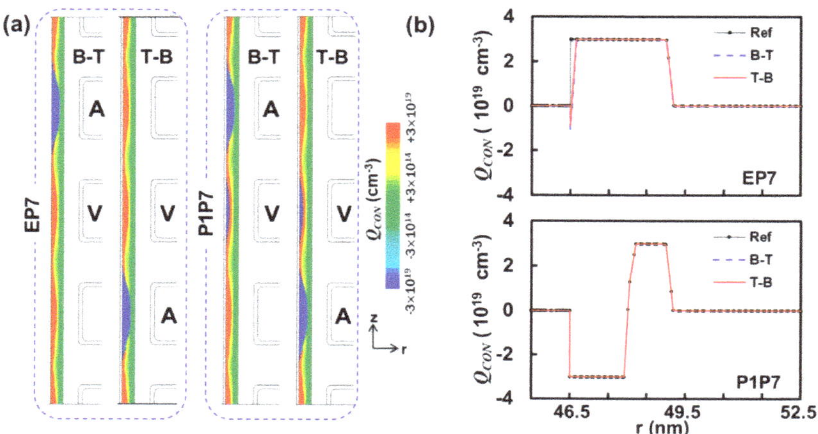

Figure 2. (**a**) Net charge concentration (Q_{CON}) comparison of B-T and T-B. (**b**) Q_{CON} in the trap-nitride layer of WL3 as a function radial coordinate r, where 46.5 nm and 52.5 nm represent two interfaces with oxide layers: top, EP7, where WL3 is originally at the state with $V_t = -3.889$ V (E). Slight change in Q_{CON} for $46.5 < \text{r (nm)} < 46.7$ is observed after interference, because of pass disturb (8 V); bottom, P1P7, where WL3 is initially at the state with $V_t = -0.976$ V (P1). In this case, pass disturb is negligible because P1 state is relatively invulnerable to $V_{PASS} = 8$ V.

The first two color plots in Figure 2a show the comparison between B-T, where WL4 was programmed to P7, and T-B, where WL2 was programmed to P7, so that all other cells appear to be similar with the peak net charge concentration of 3×10^{19} cm^{-3} except for aggressor cells with -3×10^{19} cm^{-3}. Note that the trap concentration, both for electrons and holes, was set to 3×10^{19} cm^{-3} in this work. Even though only five WLs were built in our simulation model, considering the computational cost, there was no detectable amount of asymmetry between the cell near the top (WL4) and the cell near the bottom (WL2) in terms of the net charge concentration. The plot of Q_{CON} as a function of r in Figure 2b reveals the subtle change in the net charge concentration after the aggressor cell (WL4 or WL2) is written, especially near the interface of trap-nitride and tunneling oxide ($46.5 < r$ (nm) < 46.7). The integration of the net charge concentration, with respect to the volume, led to ΔQ about -20, where B-T and T-B showed negligible difference. The color plots for P1P7 on the right in Figure 2a show P1P7 interference where the victim is programmed to P1 before the aggressor is written to P7, so that slight blue color region ($Q_{CON} < 0$) is identified together with the trapped holes from the block erasing operation. The plot of Q_{CON} versus r in Figure 2b for P1P7 demonstrates the coexistence of trapped electrons near the interface with tunneling oxide ($46.5 < r$ (nm) < 48) and trapped holes farther away from the interface ($48 < r$ (nm) < 49). More importantly, all three curves (Ref, B-T, and T-B) are almost overlapped and the corresponding integration of the difference concluded that the charge equivalent to just one electron tunneled through the victim cell under $V_{PASS} = 8$ V for 100 µs. In order to explain the sudden jump in interference from P1 to E, the information of the change in the net charge can be utilized. The upper bound of the V_t shift, as a result of the additional 20 trapped electrons, can be estimated by $\Delta V_t = 1.6 \times 10^{-19} \times \Delta Q / C$ with the assumption of a simple one-dimensional capacitor. C was calculated to be 20.6 aF by $C^{-1} = C_{\text{TOX}}{}^{-1} + C_{\text{TrapN}}{}^{-1} + C_{\text{BOX}}{}^{-1}$, and results in $\Delta V_t = 155$ mV by $\Delta Q = 3.28$ aC (20 electrons), whereas for P1P7 interference the contribution of ΔQ to interference is just 8 mV because only one electron was additionally trapped. Therefore, the distinctively high interference for EP7 should be attributed to the tunneling under $V_{PASS} = 8$ V for 100 µs, whereas P1P7 allows negligible tunneling under the same condition.

Figure 3a,b shows the band diagram for WL3 along with the radial direction from r = 35 nm (interface between poly-silicon channel and fill-oxide) to *r* = 65 nm (tungsten gate) for the aforementioned cases, EP7 and P1P7. Due to the lower conduction band edge (or electrostatic potential) of the trap-nitride layer stemming from the trapped hole (792 holes trapped after block erasing shown in Figure 3c), the tunneling barrier from the conduction band edge of the channel is partially Fowler–Nordheim type. As a result, the conduction band edge's up-lift of about 0.03 eV can be observed at *t* = 100 µs, compared to *t* = 1 µs on the inset. However, P1P7 in Figure 3b exhibits a harsher tunneling barrier because P1 state possesses only 201 holes as shown in Figure 3e; hence, the electrostatic potential of Si_3N_4 is relatively higher than that of the state E. The inset shows negligible change in the conduction band edge during 100 µs, which is consistent to the statement for P1P7 of Figure 2b (only 1 electron tunneled). Figure 3c shows the change in the number of net charges in the trap-nitride layer of WL3 as a function of time. The aggressor under 16 V shows nonlinearly fast electron tunneling as a function time, where 807 holes initially located in WL2's trap-layer are almost cancelled to neutral within 1 µs and, for the rest of the time, the additional charge corresponding to 918 electrons is trapped until *t* = 100 µs. Figure 3d,e show the change with time for EP7 and P1P7, respectively. Because the range of change is significantly small (EP7: 20 electrons and P1P7: 1 electron) compared to the aggressor cell at a larger bias of 16 V, the time-dependent evolution appears to be simple linear evolutions.

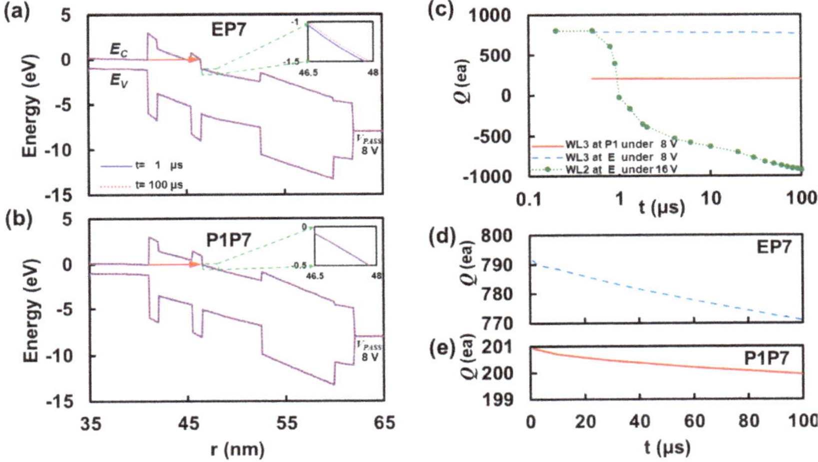

Figure 3. Band diagram of the victim cell (WL3) along the radial direction of a cylindrical cell string and corresponding number of trapped charges in the trap-nitride layer as a function of time while programming WL2 with V_{PGM} = 16 V and V_{PASS} = 8 V. (**a**) WL3 at the state E exhibits Fowler–Nordheim tunneling due to lowered conduction band edge by trapped hole charges in the trap-nitride layer. (**b**) WL3 at the state P1 depicts the harsher tunneling barrier compared to that of E in Figure 4a. This is because the net charge in the trap-nitride layer is less positive compared to E(erase) state so that the electrostatic potential is higher. (**c**) Number of trapped charges (*Q*) in the trap-nitride layer of WL2 beginning from the state E as a function of time under programming voltage V_{PGM} = 16 V is shown (green dotted line) together with that of victim cell under two different states (E and P1). (**d**) WL3 at E under the bias V_{PASS} = 8 V shows the charge in *Q* from +792 to +771, implying about 20 electrons were tunneled and holes were canceled. (**e**) WL3 at P1 shows negligible change in *Q* (from +201 to +200) so that the interference (121 mV for B-T and 88 mV for T-B) purely comes from the adjacent cell's channel inversion.

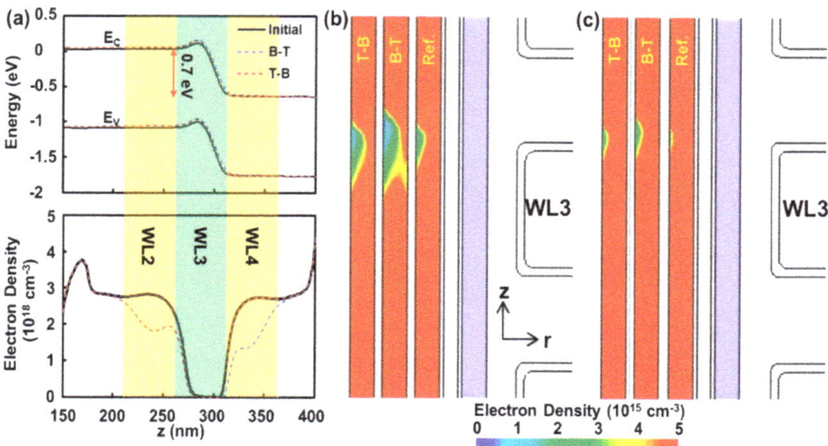

Figure 4. Poly-silicon channel information during reading operation (V_{WL3} = −1 V, $\boldsymbol{V_{READ}}$ = 7 V, V_{BL} = 0.7 V, V_{CSL} = 0 V) (**a**) band diagram: Top, electron carrier concentration; bottom, following the z-axis (r = 38 nm). The potential of 0.7 V through the bit line is mainly applied to the reading cell (WL3) since the adjacent cells are fully inverted with $\boldsymbol{V_{READ}}$ = 7 V; hence they have negligible resistances. Consequently, WL4 should experience less inversion (by $\boldsymbol{V_{READ}}$ − 0.7 V = 6.3 V) compared to WL2 (by $\boldsymbol{V_{READ}}$ − 0 V = 7 V), which is reflected in electron density in the bottom figure. WL4 and WL2 have carriers of 1.3 and 1.9 (10^{18} cm^{-3}) at the center, respectively. (**b**) Color plots of electron density for 'initial' reveal non-centered carrier bottleneck due to drain-induced-barrier-lowering (DIBL) effect. As a result, B-T; having the upper adjacent cell programmed, has the stronger interference compared to T-B with the lower adjacent cell programmed. The light blue region corresponds to trap-nitride layer (Si_3N_4) (**c**) When $\boldsymbol{V_{READ}}$ is increased to 8 V, the imbalance between B-T ($\boldsymbol{V_{READ}}$ − 0.7 V = 7.3 V) and T-B ($\boldsymbol{V_{READ}}$ − 0 = 8 V) is reduced.

Now, P1P7 can be regarded as the best example to investigate the mechanism of improved interference performance for T-B over B-T because it allows us to rule out ΔQ even after experiencing $\boldsymbol{V_{PASS}}$ = 8 V for 100 µs (pass disturb), whereas the contrast is the largest among others: P2P7, P3P7, ..., P7P7. Figure 4a shows the P1P7 case's band diagram for poly-silicon channel through the axial direction z when $\boldsymbol{V_{WL3}}$ = −1 V, which is approximately the $\boldsymbol{V_t}$ of P1 state (−0.99 V), is being applied on WL3 and $\boldsymbol{V_{READ}}$ = 7 V for the other cells. Due to the partial inversion of WL3 with −1 V compared to WL2 and WL4 with 7 V, the voltage applied to BL (V_{BL} = 0.7 V) is mainly applied to solely WL3. As a result, the upper cells, including WL4, should encounter drain-induced-barrier-lowering (DIBL), hence the actual potential drop across ONO should be 6.3 V ($\boldsymbol{V_{READ}}$ − $\boldsymbol{V_{BL}}$). The plot at the bottom of Figure 4a reveals the electron carrier density, which shows the slightly lower carrier concentration for the WL4 region compared to that for the WL2 region. Moreover, the minimum carrier concentration 5.1×10^{15} cm^{-3} appeared at z = 303 nm, which is above the center of WL3 (z = 287.5 nm) and reflects the effect of DIBL. Figure 4b visualizes the off-centered 'bottleneck' for conduction. It should be noted that the red-colored region represents that the carrier density is equal to or higher than 10^{15} cm^{-3}. Due to the off-centered bottleneck based on DIBL, the aggressor on the upper adjacent cell (WL4 for B-T case) strengthens the bottleneck which reflects high interference (121 mV in Figure 1c). For T-B case, the bottleneck is less affected by the aggressor at the lower cell (WL2) so that the interference is reduced significantly (88 mV in Figure 1c). Figure 4c shows a similar comparison under higher read voltage, $\boldsymbol{V_{READ}}$ = 8 V. Considering that the contrast between T-B and B-T comes from the DIBL effect on $\boldsymbol{V_{READ}}$, it is observed that the higher reading voltage lessens the difference between the two.

It is worth inspecting the trend of T-B compared to B-T under various circumstances and scaled cell dimensions that are inevitable for the next generation of products with more layers; unless semiconductor process hurdles related to vertical NAND's stack height are dramatically resolved, such as high aspect ratio etching technique and mechanical stress issues, to name a few [28].

Figure 5a shows the variation with respect to read voltage difference. It can be seen that the improvement by T-B over B-T is protruding with smaller read voltage such that V_{READ} = 6 V shows the improvement of 28 mV (= 107 mV − 79 mV), whereas V_{READ} = 8 V exhibits 19 mV (= 113 mV − 94 mV) when considering the averaged value of P1P7, P2P7, ..., P6P7, and P7P7. Figure 5b,c depict the trend with scaled dimensions where 24 nm for the thickness of the nitride pad during the initial stage of the VNAND process (L_g) and 17 nm for the thickness of the oxide pad (L_s) are highly probable for the newest vertical NAND Flash Memory product (>170 layers) under development. It is clearly shown that the scaled cells undergo significant interference such that L_g = 24 nm shows 162 mV and L_s = 17 nm shows 155 mV under B-T. Luckily the remedy by T-B over B-T also increases with scaling such that L_g/L_s = 24 nm/20 nm shows the improvement of 36 mV, which is superior than 21 mV from the reference geometry of this work (L_g/L_s = 31 nm/20 nm) so that the deterioration in interference and read window can be slowed down. It is noted that we simulated thicker ONO and Poly-Si cases (7.8/6.3/5.8/6.6 nm) compared to the reference (7.5/6.0/5.5/6.0 nm) to confirm any remarkable deviation owing to the stack-coverage. Nevertheless, the interference for T-B and B-T were found to be 114 mV and 86 mV, respectively, such as the reference of 105 mV for T-B and 84 mV for B-T. Therefore, we believe that the state-of-the-art stack coverage (ONO > 95% and Poly-Si > 90%) in the Flash memory product's thin-film process is sufficiently good enough to impose little uncertainties in our simplified TCAD models.

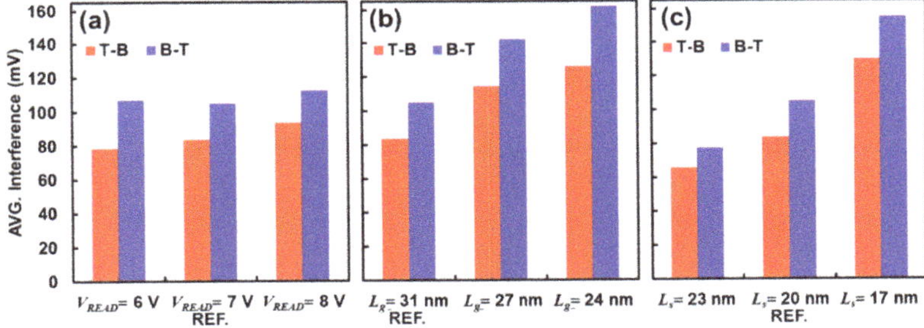

Figure 5. Averaged interference (P1P7, P2P7, ..., P6P7, and P7P7) for T-B and B-T schemes with various changes such as (**a**) V_{READ}, and cell dimensions, (**b**) gate length and (**c**) gate space, for the next generation vertical NAND Flash products. Note that the reference is (V_{READ}, L_g, L_s) = (7 V, 31 nm, 20 nm) and the raw data of each case is available in Figure S1.

Figure 6 exposes the corresponding electrostatic potential distribution for L_g = 24 nm compared to the reference L_g = 31 nm, further analyzing the improvement by T-B scheme for scaled cells as an example. It should be noted that the electrostatic potential is referenced to that of WL3. P1P7 interference, where $V_{t,\mathrm{WL3}}$ = −0.99 V and $V_{t,\mathrm{Aggressor}}$ = 3.18 V, was used consistently for the analysis in Figure 4, which exhibits the improvement from 121 mV of B-T to 88 mV of T-B as shown in Figure 1c. Note that the case of L_g = 24 nm makes 234 mV from B-T and 165 mV from T-B, which is higher than the averaged values available in Figure 5. The electrostatic potential valley is mainly responsible for the V_t of the cell under reading, and it is observed that at the center of the channel (r = 38 nm) the length of the valley (0.4 V < electrostatic potential < 0.47 V) changes dramatically for the scaled cell (15 nm → 12 nm at L_g = 24 nm), compared to the reference (19 nm → 17 nm at L_g = 31 nm). The emphasized deterioration in interference with scaled NAND cell sizes is indirectly evidenced by a 14 nm planar NAND flash memory reported in 2016 by Samsung [23].

Although they did not adopt the scheme of T-B [22] and kept the conventional B-T due to undisclosed reasons, a significant interference (back pattern dependency or back-pattern-effect) in the extremely scaled 14 nm NAND cells might have forced them to introduce a new scheme in incremental step pulse programming (ISPP), where V_{READ} is lowered selectively for upper cells during the verify operation in ISPP.

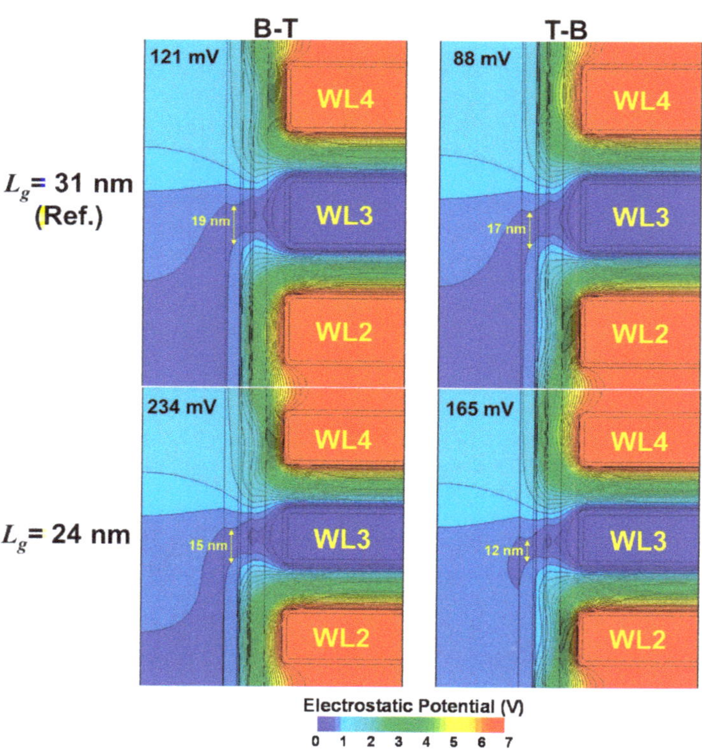

Figure 6. Electrostatic potential distribution change with L_g scaling (31 → 24 nm) Both are after EP7 interference ($V_{t,Victim}$ = −0.99 V, $V_{t,Aggressor}$ = 3.2 V) followed by reading at the moment at V_{WL3} = −1 V.

4. Conclusions

In conclusion, this work performed a systematic study on the improvement in interference when the Top to Bottom (T-B) programming scheme is employed compared to the conventional Bottom to Top (B-T) scheme which probably originated from the planar NAND Flash products with a single layer on the ground level in a historical point of view. With the aid of TCAD simulations, it is shown that only the erased state (E) suffers from both pass disturb under the normal condition of V_{PASS} = 8 V and coupling to the adjacent cells. The enhancement by the T-B scheme is mainly delivered by the latter contribution (coupling), stemming from the nature of NAND's reading operation combined with drain-induced-barrier-lowering (DIBL). Therefore, most states (e.g., P1, P2, ... , P6, P7 for TLC and P1, P2, ... , P14, P15 for QLC) can benefit from the T-B scheme, despite the fact that programmed states are inherently free from pass disturb. Moreover, it is expected that T-B lessens the interference more prominently, especially for the next generation vertical NAND Flash products with more than 170 layers, inevitably followed by the higher degree of integration (smaller L_g and L_s). This work highlights its importance for future vertical NAND Flash memories, the applications of which include conventional use as data storage [21], but also other applications such as neuromorphic computing [29–32], security in IoTs [33], etc.

Supplementary Materials: The supporting materials are available online at https://www.mdpi.com/article/10.3390/mi12050584/s1. Figure S1: Raw data of interference under various conditions of V_{READ}, L_g, and L_s with respect to the reference V_{READ} = 7 V, L_g = 31 nm, and L_s = 20 nm shown in Figure 5.

Author Contributions: Conceptualization, methodology, and original draft preparation, S.-i.Y. Validation, funding acquisition, and writing—review and editing, J.K. Both authors have read and agreed to the published version of the manuscript.

Funding: This work was supported by the National Research Foundation of Korea (NRF) grant funded by the Korea government (MSIT) (No. 2020R1G1A1099554).

Acknowledgments: The EDA Tool was supported by the IC Design Education Center (IDEC).

Conflicts of Interest: The authors declare no conflict of interest.

References

1. Mahmoodi, M.R.; Prezioso, M.; Strukov, D.B. Versatile stochastic dot product circuits based on nonvolatile memories for high performance neurocomputing and neurooptimization. *Nat. Commun.* **2019**, *11*, 5113. [CrossRef] [PubMed]
2. Guo, X.; Bayat, F.M.; Bavandpour, M.; Klachko, M.; Mahmoodi, M.R.; Prezioso, M.; Likaharev, K.K.; Strukov, D.B. Fast, energy-efficient, robust, and reproducible mixed-signal neuromorphic classifier based on embedded NOR flash memory technology. In Proceedings of the 2017 IEEE International Electron Devices Meeting (IEDM), San Francisco, CA, USA, 2–6 December 2017; pp. 1–4.
3. Guo, X.; Bayat, F.M.; Prezioso, M.; Chen, Y.; Nguyen, B.; Do, N.; Strukov, D.B. Temperature-insensitive analog vector-by-matrix multiplier based on 55 nm NOR flash memory cells. In Proceedings of the 2017 IEEE Custom Integrated Circuits Conference (CICC), Austin, TX, USA, 30 April–3 May 2017; pp. 1–4.
4. Jang, J.; Kim, H.-S.; Cho, W.; Cho, H.; Kim, J.; Shim, S.-I.; Jang, Y.; Jeong, J.-H.; Son, B.-K.; Kim, D.W.; et al. Vertical Cell Array using TCAT (Terabit Cell Array Transistor) Technology for Ultra High Density NAND Flash Memory. In Proceedings of the 2009 IEEE Symposium on VLSI Technology, Kyoto, Japan, 15–17 June 2009.
5. Compagnoni, C.M.; Goda, A.; Spinelli, A.S.; Feeley, P.; Lacaita, A.L.; Visconti, A. Reviewing the Evolution of the NAND Flash Technology. *Proc. IEEE* **2017**, *105*, 1609–1633. [CrossRef]
6. Choe, B.-I.; Lee, J.-K.; Park, B.-G.; Lee, J.-H. Suppression of Read Disturb Fail Caused by Boosting Hot Carrier Injection Effect for 3-D Stack NAND Flash Memories. *IEEE Electron Dev. Lett.* **2014**, *35*, 42–44. [CrossRef]
7. Kwon, D.W.; Lee, J.; Lee, R.; Kim, S.; Lee, J.-H.; Park, B.-G. Novel Boosting Scheme Using Asymmetric Pass Voltage for Reducing Program Disturbance in 3D NAND Flash Memory. *IEEE J. Electron Dev. Soc.* **2018**, *6*, 286–290. [CrossRef]
8. Li, Q.; Shi, L.; Di, Y.; Du, Y.; Xue, C.J.; Yang, C.; Zhuge, Q.; Sha, E.H.M. Improving read performance via selective Vpass reduction on high density 3D NAND flash memory. In Proceedings of the 2017 IEEE 6th Non-Volatile Memory Systems and Applications Symposium (NVMSA), Hsinchu, Taiwan, 16–18 August 2017; pp. 1–4.
9. Zhang, Y.; Jin, L.; Jiang, D.; Zou, X.; Liu, H.; Huo, Z. A Novel Read Scheme for Read Disturbance Suppression in 3D NAND Flash Memory. *IEEE Electron Dev. Lett.* **2017**, *38*, 1669–1672. [CrossRef]
10. Flash Memory Summit. Available online: https://flashmemorysummit.com/ (accessed on 17 May 2020).
11. Han, C.; Wu, Z.; Yang, C.; Xie, L.; Xu, B.; Liu, L.; Liu, L.; Yin, Z.; Jin, L.; Huo, Z. Influence of accumulated charges on deep trench etch process in 3D NAND memory. *Semicon. Sci. Technol.* **2020**, *35*, 045003. [CrossRef]
12. Neumann, J.T.; Klochkov, D.; Korb, T.; Gupta, S.; Avishai, A.; Pichumani, R.; Lee, K.; Buxbaum, A.; Foca, E. 3D analysis of high-aspect ratio features in 3D-NAND. *Proc. SPIE* **2020**, *11325*, 1–11.
13. Ye, Y.; Xia, Z.; Liu, L.; Huo, Z. Investigation of Reducing Bow during High Aspect Ratio Trench Etching in 3D NAND Flash Memory. In Proceedings of the 2018 14th IEEE International Conference on Solid-State and Integrated Circuit Technology (ICSICT), Qingdao, China, 31 October–3 November 2018; pp. 1–3.
14. Choe, H.; Jee, J.; Lim, S.C.; Joe, S.M.; Park, I.H.; Park, H. Machine-Learning-Based Read Reference Voltage Estimation for NAND Flash Memory Systems Without Knowledge of Retention Time. *IEEE Access* **2020**, *8*, 176416–176429. [CrossRef]
15. Ko, K.; Lee, J.K.; Shin, H. Variability-Aware Machine Learning Strategy for 3-D NAND Flash Memories. *IEEE Trans. Electron Dev.* **2020**, *67*, 1575–1580. [CrossRef]
16. Kang, D.; Jeong, W.; Kim, C.; Kim, D.-H.; Cho, Y.-S.; Kang, K.-T.; Ryu, J.; Kang, K.-M.; Lee, S.; Kim, W.; et al. 256 Gb 3b/cell V-NAND Flash Memory with 48 Stacked WL Layers. *IEEE J. Solid State Circuits* **2017**, *52*, 210–217. [CrossRef]
17. Kim, C.; Cho, J.-H.; Jeong, W.; Park, I.-H.; Park, H.-W.; Kim, D.-H.; Kang, D.; Lee, S.; Lee, J.-S.; Kim, W.; et al. A 512Gb 3b/cell 64-Stacked WL 3D V-NAND Flash Memory. In Proceedings of the 2017 IEEE International Solid-State Circuits Conference, San Francisco, CA, USA, 11–15 February 2017.

18. Kim, C.; Kim, D.-H.; Jeong, W.; Kim, H.-J.; Park, I.-H.; Park, H.-W.; Lee, J.-H.; Park, J.-Y.; Ahn, Y.-L.; Lee, J.-Y.; et al. 512-Gb 3-b/Cell 64-Stacked WL 3-D-NAND Flash Memory. *IEEE J. Solid State Circuits* **2018**, *53*, 124–133. [CrossRef]
19. Lee, S.; Kim, C.; Kim, M.; Joe, S.-M.; Jang, J.; Kim, S.; Lee, K.; Kim, J.; Park, J.; Lee, H.-J.; et al. A 1Tb 4b/Cell 64-Stacked-WL 3D NAND Flash Memory with 12MB/s Program Throughput. In Proceedings of the 2018 IEEE International Solid-State Circuits Conference, San Francisco, CA, USA, 11–15 February 2018.
20. Maejima, H.; Kanda, K.; Fujimura, S.; Takagiwa, T.; Ozawa, S.; Sato, J.; Shindo, Y.; Sato, M.; Kanagawa, N.; Musha, J.; et al. A 512Gb 3b/Cell 3D Flash Memory on a 96-Word-Line-Layer Technology. In Proceedings of the 2018 IEEE International Solid-State Circuits Conference, San Francisco, CA, USA, 11–15 February 2018.
21. Kang, D.; Kim, M.; Jeong, S.-C.; Jung, W.; Park, J.; Choo, G.; Shim, D.-K.; Kavala, A.; Kim, S.-B.; Kang, K.-M.; et al. A 512Gb 3-bit/Cell 3D 6th-Generation V-NAND Flash Memory with 82MB/s Write throughput and 1.2Gb/s Interface. In Proceedings of the 2019 IEEE International Solid-State Circuits Conference, San Francisco, CA, USA, 17–21 February 2019.
22. Chen, W.-C.; Lue, H.-T.; Chang, K.-P.; Hsiao, Y.-H.; Hsieh, C.-C.; Shih, Y.-H.; Lu, C.-Y. Study of the Programming Sequence Induced Back-Pattern Effect in Split-Page 3D Vertical-Gate (VG) NAND Flash. In Proceedings of the 2014 IEEE International Symposium on VLSI Technology Systems and Application, Hsinchu, Taiwan, 28–30 April 2014.
23. Lee, S.; Lee, J.-Y.; Park, I.-H.; Park, J.; Yun, S.-W.; Kim, M.-S.; Lee, J.-H.; Kim, M.; Lee, K.; Kim, T.; et al. A 128Gb 2b/cell NAND Flash memory in 14nm Technology with t_{PROG}=640µs and 800MB/s I/O Rate. In Proceedings of the 2019 IEEE International Solid-State Circuits Conference, San Francisco, CA, USA, 17–21 February 2019.
24. Synopsys. Sentaurus Manual S-Device; San Jose, CA, USA, L.-Version. 2016. Available online: https://www.synopsys.com/silicon/tcad/device-simulation/sentaurus-device.html/ (accessed on 17 May 2020).
25. Nowak, E.; Kim, J.-H.; Kwon, H.-Y.; Kim, Y.-G.; Sim, J.-S.; Lim, S.-H.; Kim, D.-S.; Lee, K.-H.; Park, Y.-K.; Choi, J.-H.; et al. Intrinsic Fluctuations in Vertical NAND Flash Memories. In Proceedings of the 2012 Symposium on VLSI Technology, Honolulu, HI, USA, 12–14 June 2012.
26. Oh, Y.-T.; Kim, K.-B.; Shin, S.-H.; Sim, H.; Toan, N.V.; Ono, T.; Song, Y.-H. Impact of etch angles on cell characteristics in 3D NAND flash memory. *Microelectron. J.* **2018**, *79*, 1–6. [CrossRef]
27. Lee, J.-K.; Ko, K.; Shin, H. Analysis on Process Variation Effect of 3D NAND Flash Memory Cell through Machine Learning Model. In Proceedings of the 4th IEEE Electron Devices Technology & Manufacturing Conference (EDTM), Penang, Malaysia, 16–18 March 2020.
28. Kim, H.; Ahn, S.; Shin, Y.G.; Lee, K.; Jung, E. Evolution of NAND Flash Memory: From 2D to 3D as a Storage Market Leader. In Proceedings of the 2017 IEEE International Memory Workshop (IMW), Monterey, CA, USA, 14–17 May 2017; pp. 1–4.
29. Shim, W.; Yu, S. Technological Design of 3D NAND-Based Compute-in-Memory Architecture for GB-Scale Deep Neural Network. *IEEE Electron Dev. Lett.* **2021**, *42*, 160–163. [CrossRef]
30. Bavandpour, M.; Sahay, S.; Mahmoodi, M.R.; Strukov, D.B. 3D-aCortex: An ultra-compact energy-efficient neurocomputing platform based on commercial 3D-NAND flash memories. *arXiv* **2019**, arXiv:1908.02472.
31. Xiao, T.P.; Bennett, C.H.; Feinberg, B.; Agarwal, S.; Marinella, M.J. Analog architectures for neural network acceleration based on non-volatile memory. *Appl. Phys. Rev.* **2020**, *7*, 031301. [CrossRef]
32. Yi, S.-I.; Kumar, S.; Williams, R.S. Improved Hopfield Network Optimization using Manufacturable Three-terminal Electronic Synapses. *arXiv* **2021**, arXiv:2104.12288.
33. Larimian, S.; Mahmoodi, M.R.; Strukov, D.B. Lightweight Integrated Design of PUF and TRNG Security Primitives Based on eFlash Memory in 55-nm CMOS. *IEEE Trans. Electron Dev.* **2020**, *67*, 1586–1592. [CrossRef]

micromachines

MDPI

Review

Random Telegraph Noise in 3D NAND Flash Memories

Alessandro S. Spinelli *, Gerardo Malavena, Andrea L. Lacaita and Christian Monzio Compagnoni

Dipartimento di Elettronica, Informazione e Bioingegneria, Politecnico di Milano, 20133 Milan, Italy; gerardo.malavena@polimi.it (G.M.); andrea.lacaita@polimi.it (A.L.L.); christian.monzio@polimi.it (C.M.C.)
* Correspondence: alessandro.spinelli@polimi.it; Tel.: +39-02-2399-4001

Abstract: In this paper, we review the phenomenology of random telegraph noise (RTN) in 3D NAND Flash arrays. The main features of such arrays resulting from their mainstream integration scheme are first discussed, pointing out the relevant role played by the polycrystalline nature of the string silicon channels on current transport. Starting from that, experimental data for RTN in 3D arrays are presented and explained via theoretical and simulation models. The attention is drawn, in particular, to the changes in the RTN dependences on the array working conditions that resulted from the transition from planar to 3D architectures. Such changes are explained by considering the impact of highly-defective grain boundaries on percolative current transport in cell channels in combination with the localized nature of the RTN traps.

Keywords: 3D NAND Flash memories; random telegraph noise; Flash memory reliability

Citation: Spinelli, A.S.; Malavena, G.; Lacaita, A.L.; Monzio Compagnoni, C. Random Telegraph Noise in 3D NAND Flash Memories. *Micromachines* **2021**, *12*, 703. https://doi.org/10.3390/mi12060703

Academic Editors: Cristian Zambelli and Rino Micheloni

Received: 11 May 2021
Accepted: 4 June 2021
Published: 16 June 2021

1. Introduction

Random telegraph noise (RTN) in MOS transistors has been an important topic of interest in the solid-state device community since the 80s, when results of low-frequency noise characterization [1] showed a transition from a typical $1/f$ behavior at high temperatures to a series of discrete switching events as temperature was lowered. Similar observations were soon made when moving from large- to small-area devices [2], and interpreted in terms of capture/emission of electrons by single interface traps. On the theoretical side, this result highlighted the importance of the number fluctuation contribution to the flicker noise, but prompted the emergence of a new limitation to MOS device operation as well [3].

Moving from early investigations and models [4–9], the RTN picture grew more complex, as novel time and amplitude observations [10–14] hinted at a non-negligible role played by non-uniform electron conduction in submicron devices [15,16]. This idea gained traction when the phenomenon began to be investigated in Flash memories [17–23], demonstrating current fluctuations up to 60% [22] and threshold voltage (V_T) shifts reaching 700 mV [18] in 90-nm technology node devices. The physical picture now accepted that accounts for such results is based on the fact that, in scaled devices, dopants must be viewed as individual ions rather than a continuous distribution, resulting in randomly-placed charges in the depletion region. Such random point charges [24–27] give rise to sharp peaks in the band energy profile of the channel of an MOS transistor, resulting in local modulation of the current flow and filamentary conduction. If a "strategic" trap happens to be placed right above a current path, electron trapping will effectively shut off such a path, resulting in a large drain current and V_T fluctuation [28–31]. On the other hand, if the trap is placed over a region in which little current flows, its trapping/detrapping will barely affect the overall current. Such an idea has been successfully applied to explain the statistical distribution of the RTN fluctuations in NOR and NAND arrays, measured in terms of their amplitude [19,32,33] and time constants [34–36], providing a useful tool for extracting information about the impact of device parameters on RTN. A recent review of the issue can be found in [37].

The above-mentioned framework has served nicely the Flash community until the first decade of the 21st century, when several limitations to the scaling of the planar NAND technology prompted the emergence of 3D arrays [38]. In such devices, the RTN picture just outlined fell short of adequately describing the experimental data, in view of the peculiar characteristics of the polycrystalline material used as conduction channel.

In the following, we will review in detail the physics of RTN in 3D NAND Flash memories, discussing the main experimental data and physical models developed to quantitatively account for them. We begin our discussion with a brief summary of the main 3D array architecture and cell structure, followed by a description of electron transport in 3D NAND channels. This will allow us to develop a consistent picture of RTN in 3D NAND devices, whose main features will be highlighted. After this part, we will focus our attention on the main experimental data presented in the literature, taking advantage of the model results to provide interpretation for them.

2. Array and Cell Structure

Among the several architectural solutions for 3D storage [39–46], the one employing vertical-channel strings crossed by a set of planar wordlines has become the most effective one [47–50], and is the focus of this section. Here we will briefly describe the main features of such an array, namely its organization and cell structure, referring to previous works for further details [51].

A pictorial view of the array is shown in Figure 1 (left): note that the cell strings run vertically from the substrate to the bitlines. As in planar arrays, select elements are needed near the source and drain ends of the string, integrated in rows running orthogonally to the bitlines. The rest of the cells are contacted by planar wordlines that span over an entire block of the array. One of the advantages of this structure is that the large increase in density allowed by the exploitation of the third dimension makes it possible to relieve some of the pressure on channel length scaling and its many drawbacks from the viewpoint of process complexity and reliability, well known in planar devices [52,53]: cell length in 3D NAND is around 25–30 nm [54], with the additional advantage of becoming less dependent on the availability of advanced lithography tools. A second advantage of this solution lies in its manufacturing process: memory cells are not patterned individually, but they are formed all at once as cylindrical holes are cut through the stacked wordlines, creating the strings. This procedure entails that the elementary cell becomes a gate-all-around, vertical-channel transistor, with the advantage of a better electrostatic control from the gate. A schematic view of such a device is shown in Figure 1 (right): starting from the outside we meet a contacted wordline, a blocking dielectric and a charge-storage layer, that can either be a floating gate [47,55–59], similar to planar NAND devices, or a charge-trap layer [60–64], followed by the tunnel oxide. Beyond the oxide, we can notice a thin silicon region and an inner oxide filling the central region of the cylinder, labeled filler oxide for simplicity. This structure, where the conductive channel is a hollow cylinder, is referred to as a "Macaroni" MOSFET, and is the result of clever device engineering in 3D NAND: in fact, after the vertical high-aspect ratio holes have been etched in the structure of Figure 1 (left), and the blocking, storage and tunnel layers deposited, the remaining part of the cylinder must be filled with silicon. The result is a polycrystalline channel whose central region is plagued by a large defectivity, impairing the device performance. To avoid such a drawback, a very thin polysilicon layer is deposited on the gate dielectric, while the remaining central region of the cylinder is filled with a dielectric [40], gaining two distinct advantages: first, thinning of the silicon body results in reduced short-channel effects and better electrostatic control from the gate; second, defect removal further contributes to better subthreshold slope and array performance.

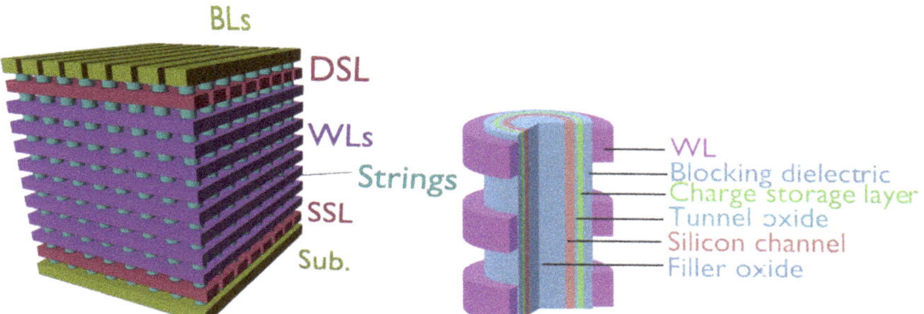

Figure 1. (**Left**) Conceptual view of a vertical-channel 3D NAND array with its main elements (SSL = source select line, WLs = wordlines, DSL = drain select line, BLs = bitlines). (**Right**) pictorial view of an array string highlighting the structure of the elementary memory cells.

3. Polysilicon Conduction

The overview on the memory cell design given in the previous section already suggested that the polycrystalline character of the conduction channel is a key parameter from the viewpoint of device performance. A polycrystalline material, in fact, is formed by single-crystal regions labeled grains, with different crystallographic orientations. Such regions are separated by highly-defective interfaces, or grain boundaries (GBs) [65]. A pictorial view of a NAND string of ten cells with its inner polysilicon region and grains is shown in Figure 2: note the random structure of grains and GBs, that are the key elements affecting device current and variability.

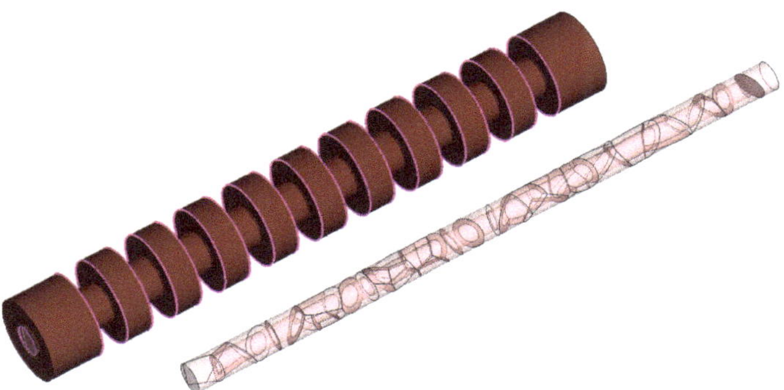

Figure 2. Pictorial view of a ten-cell memory string (**left**) and of the inner polysilicon regions separated into polycrystalline grains (**right**). The example is the result of a TCAD simulation of the cell structure where polysilicon grains are obtained via Voronoi tessellation of the silicon region.

One of the key properties of polysilicon is its trap density, whose value has been estimated by several works, based on either direct optical or electrical experimental measurements [66–74] or via numerical device simulations [75–79]. Many of such results point to a double-exponential energy distribution of donor-like and acceptor-like states of the form (for acceptor-like states in the upper half of the energy gap):

$$N_{GB}(E) = N_T e^{-(E-E_C)/E_T} + N_D e^{-(E-E_C)/E_D}, \tag{1}$$

where the reported range for the acceptor-like states parameters is listed in Table 1. Note that the first exponential distribution is characterized by a large peak density N_T and a small characteristic energy E_T, and is usually referred to as tail states distribution, as a consequence of its location near the edge of the gap. The second has a lower peak density N_D but a higher energy E_D, and is usually labeled deep states distribution. Note also that trap densities are given as volumetric densities: this was useful in early simulation works, where a uniform trap density in the semiconductor body was assumed for simplicity. From a physical viewpoint, however, traps are expected to be mainly located at GBs, and an areal density σ is then needed. A conversion between volumetric and areal densities is readily achieved assuming for simplicity a spherical grain size with radius r_G, and placing all volume traps on the sphere surface. This leads to

$$4\pi r_G^2\sigma = \frac{4}{3}\pi r_G^3 N_{GB} \Rightarrow \sigma = \frac{r_G}{3} N_{GB}, \tag{2}$$

or a very similar conversion factor as in [79].

Table 1. Range of parameter values for the acceptor-like states in the polysilicon, according to the literature (see text for references).

N_T [cm^{-3} eV^{-1}]	E_T [meV]	N_D [cm^{-3} eV^{-1}]	E_D [meV]
9×10^{19}–10^{21}	16.6–80	1.2×10^{18}–9×10^{19}	80–500

Electron transport in polysilicon has been studied since the 70s, as this material found applications in resistors, interconnections, and silicon-gate MOSFETs. From the viewpoint of current conduction, we can identify two modeling approaches, that differ in the way GBs are treated: one approach is to extend the drift-diffusion model usually adopted in monocrystalline silicon, describing GBs as trapping centers with a reduced mobility [80–82]; the other is based on a thermionic emission model at the GBs [83–87]. Although the latter seems to be gaining traction in recent literature, a definitive conclusion has not been reached, yet, and a recent study of the different dependences implied by such models can be found in [88,89].

The above-mentioned numerical models of conduction have been used to investigate the effect of GBs on variability in nanowires [90–95] and 3D NAND devices [96,97]. A recent study based on a drift-diffusion transport within the grains and thermionic emission at the GBs [98–101] has demonstrated a good capability to reproduce several features of experimental data, including its temperature dependence. Figure 3 (left) shows a typical conduction-band profile along the channel of a 3D NAND string, for increasing values of the control-gate bias, as resulting from such model. Note that the profile is not smooth, featuring peaks in correspondence of the highly-defective GBs. As gate bias is increased, the band bending lowers the conduction-band profile, increasing the localized trap occupation and sharpening the peaks, which become the true bottlenecks of conduction [100]. This result makes clear that GBs are an additional source of non-uniformity in the current conduction, which means that they might be expected to play a main role in RTN. This is even more apparent if we consider that GB trap densities (see Table 1) are much larger than typical doping concentrations used in 3D NAND strings. A similar approach was also followed by [102].

The above-mentioned model has been applied to investigate the impact of GB traps on RTN [99,101] within a Monte Carlo approach: random configuration of GBs are first generated in the silicon region after a Voronoi tessellation [92], and traps are placed at the interfaces following the previously-discussed energy distribution. Drain current is computed up to a specified threshold, defined at a constant current level, after which an additional RTN trap is filled with an electron and the resulting V_T shift computed. Results for a template device are also reported in Figure 3, for the case of a single trap placed at one random position in a GB, and for a trap placed at a random position at the silicon/gate oxide interface. It is clear that GB traps are much more effective in modulating the electron conduction and result in larger V_T fluctuations.

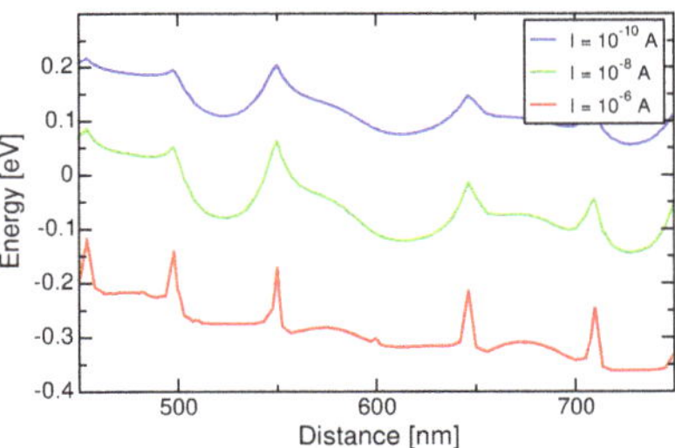

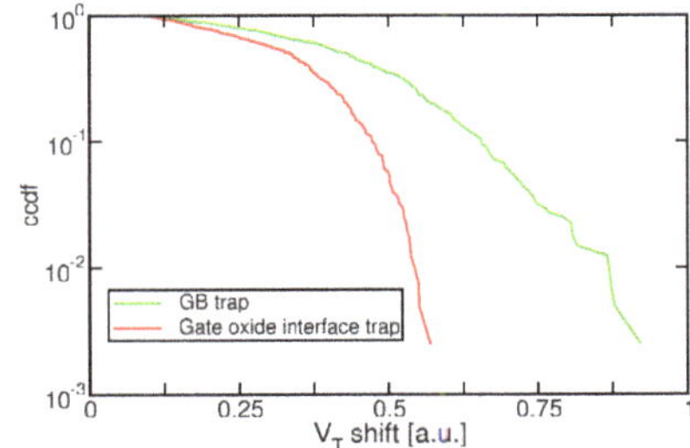

Figure 3. (**Left**) Conduction band profile for a 3D NAND string at different current levels. (**right**) RTN complementary cumulative distribution functions (ccdfs) for traps placed at the GBs or at the gate/oxide interface for a template device.

In spite of these encouraging results, several important features of this model still have to be assessed, such as the actual grain size [103–105], the mobility degradation and conduction process at the grain boundaries [106,107], and the impact of all these quantities, including architectural parameters and cell design, on RTN.

4. Experimental Data

The previous section was meant to provide a framework for the interpretation of the most relevant experimental data presented in the literature, that are discussed in the following. It must be noted, however, that RTN, as well as other reliability concerns in 3D NAND memories, remains a highly-confidential matter and very few data are published. We begin our analysis of RTN with single-trap data, moving then to statistical distributions and impact on device performance.

4.1. Single-Trap Data

Investigation of the microscopic properties of RTN single traps in 3D NAND devices can be found in [108,109], where a statistical analysis of the noise power spectral density was also carried out. In those papers it was reported that the string current fluctuations due to single-trap RTN depend on the sensing current: as the current is increased, its fluctuations also increase when measured in absolute terms, but decrease in terms of relative change. Such a dependence was also found in [110] for the above-threshold region, and ascribed to the effect of traps at the silicon-oxide interfaces. These dependences reflect similar behaviors observed both experimentally and numerically in planar or cylindrical devices [111–113], where the increased screening exerted by the mobile carriers as the gate bias is raised, mitigating the effect of the RTN trap, was invoked as an explanation. Several works reported investigations of the capture and emission time constants and their dependence on gate bias and temperature, including the activation energies [114–117]. Their results do not point to any particular difference in the microscopic nature of such traps with respect to those active in planar devices (apart from a faster capture/emission dynamics suggested in [114]): this of course is not surprising and supports an interpretation of the RTN phenomenon based on the spatial distribution of such traps rather than on some peculiar characteristics.

4.2. Array Statistical Data

From the viewpoint of the memory performance, the statistical distribution of the RTN-induced ΔV_T is the main parameter. This kind of fluctuations in poly-Si channels were first shown (to our knowledge) in [118], on a nanowire structure (no filler oxide), showing an exponential distribution for ΔV_T, which is a typical result of a percolation process. The same exponential dependence was reported on vertical NAND devices in [119–123], suggesting that the RTN distribution in arrays follows an $e^{-\Delta V_T/\lambda}$ law, and can be effectively characterized by the slope λ of the exponential distribution.

A comparison between 3D and planar cell RTN is reported in [124], where a larger RTN distribution was reported for the former, while an opposite result was claimed in [121]. It is obviously difficult if not impossible to critically assess those results and search for the reason of this discrepancy. However, from a general standpoint, the slope λ is related to both the trap density (affecting the percolation centers) and the electrostatic impact of a single trapped electron, that have an opposite trend when moving from planar to 3D devices: 3D cells are expected to have a higher trap density thanks to the presence of GBs, but feature also a larger cell (i.e., a larger capacitance and a lower electrostatic impact of a single electron). So, the different results might just be a consequence of different cell designs.

The impact of GB traps on RTN can also be noted in the comparison reported in [125,126] and carried out as a function of temperature in the range from -10 to 125 °C, that is shown in Figure 4 (left). First, please note that the shape of the two distributions is different: in 2D cells we notice clear exponential tails due to RTN departing from a central distribution, related to measurement noise in cells not affected by RTN; in 3D arrays, instead, we notice a single exponential distribution, suggesting that the large majority of cells in the 3D array are affected by RTN. A second point to stress is that the slope of the exponential distribution is reduced with respect to planar technologies [121,125,127]. Given the previous point, such an improvement seems mainly a consequence of the larger cell size of 3D arrays, although a role could also be played by the different conduction mechanism and percolation in planar and 3D devices (see for example [113] for a discussion on the RTN dependences in 3D devices). Finally, different temperature dependences are also apparent: while planar device RTN is temperature-independent [128], 3D NAND exhibit a decrease in λ at higher temperatures, as also reported in [115,116].

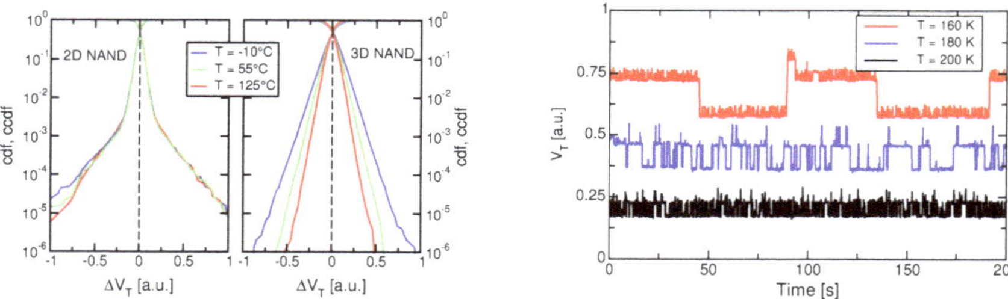

Figure 4. (**Left**) RTN cumulative density function (cdf) and its complementary (ccdf) for 2D and 3D NAND arrays at different temperatures [125,126], © 2017, IEEE. (**Right**) V_T fluctuations due to single RTN traps at different temperatures.

Such a different temperature dependence is important from a reliability standpoint and deserves further investigation. To this aim, the right side of Figure 4 shows the behavior of a single RTN trap as a function of time, for different temperatures. Besides a decrease in the absolute value of V_T for higher temperatures, reflecting an increase in the current, it is obvious that the fluctuation amplitude is decreasing as well. This behavior has been observed on a number of traps [126] and is the responsible for the improved RTN distribution. At first glance, the temperature dependence could be simply related to the thermal energy of the electrons and their better or worse capability to overcome the energy barriers, but this would not explain the difference between planar and 3D dependences. So, we must assume that temperature affects the percolation itself. To check this, we conducted simulations with the model presented in the previous section [98–101], for a template 3D NAND device at different temperatures. Results are presented in Figure 5 (left). Note that the decrease of the RTN slope at higher temperature is accounted for by the model, allowing to exploit its results to provide some more insight: to this aim, we have simulated a template device with a single GB orthogonal to the current flow and placed at the middle of the gate. Results for the conduction band at threshold at different temperatures are reported in Figure 5 (right), and feature significant differences: indeed, the conduction band peak, located at the GB and due to the localized trapped charge, becomes sharper at low temperatures, meaning that there is an increased trapped charge at low temperatures, resulting in more percolation centers. A reason for this lies in our definition of the threshold condition, that is a constant-current (10 nA) criterion. When temperature is lowered, thermal emission is reduced, and the string current lowers. To reach the same 10 nA value, gate bias must be increased, lowering the conduction band and leading to additional trap filling. Note also that this phenomenon does not take place in planar devices, where the percolation centers are the ionized dopants, whose density obviously does not change with the gate bias.

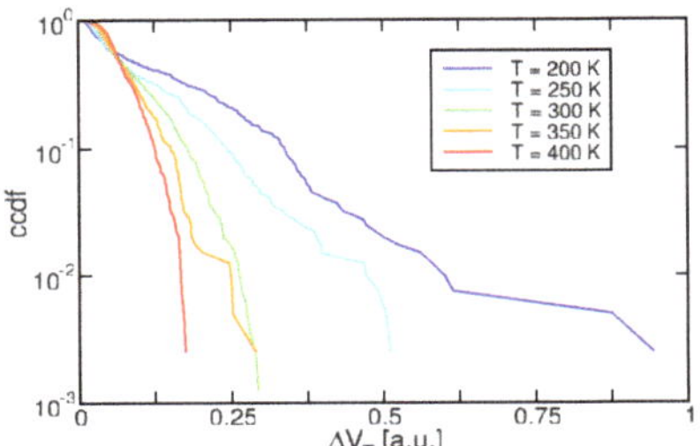

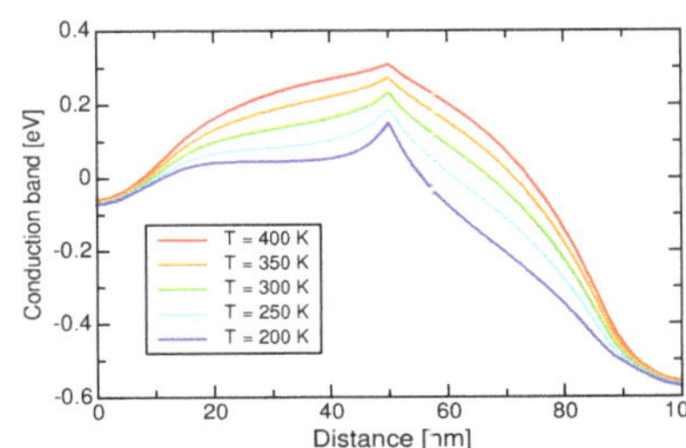

Figure 5. (**Left**) Simulation results for the RTN ccdf in a template 3D NAND device for different temperatures. (**Right**) Conduction band profile at different temperatures for a template device with a single GB located at the center of the channel.

Additional dependences exhibited by random telegraph noise in 3D NAND were reported by [108,122] with reference to read current and pass voltage. Figure 6 (left) shows RTN data for cells on different wordlines as a function of the read current and pass voltage. While some cells do not exhibit significant RTN, the one labelled as WL2 features a decreasing relative fluctuation of the current as the read current is increased, in agreement with data previously discussed. However, data also show a dependence on the pass voltage, whose increase leads to a higher RTN. Similar data are reported in Figure 6 (right), where the RTN distribution is shown. Data show that the tail slope of the bitline current increases as the pass voltage is increased. This result was related to the previous one by the authors of [122], as increasing the pass voltage means a reduction of the read threshold voltage and an increase in the RTN fluctuations. However, further analysis are needed to clarify the link between the string operating conditions and the measured RTN.

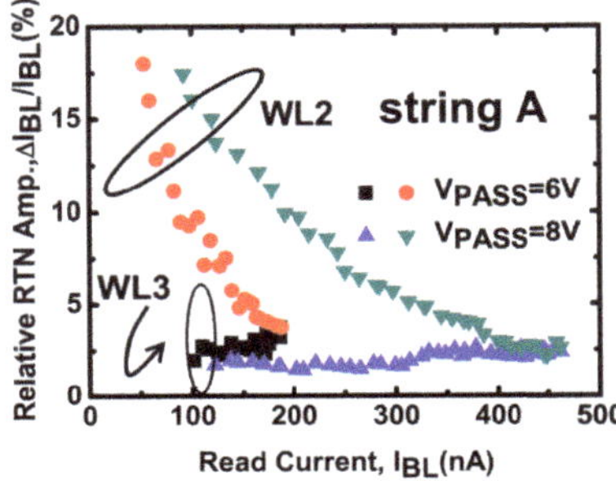

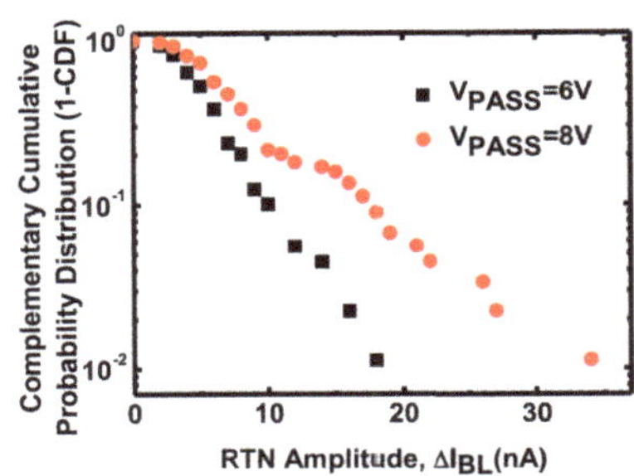

Figure 6. RTN ccdf as a function of the read current (**left**) and of the pass bias, at a read current of 100 nA (**right**). From [122], © 2016, IEEE.

Finally, the effect of cycling on RTN in 3D arrays has been investigated in [110,119,126] and data from [126] are reported in Figure 7 (left). Note that the RTN ΔV_T data for a fresh and a cycled array show only a minimal increase in the height and slope of the distribution, which is again different from the noticeable increase in the RTN distribution reported in planar devices (see [128,129]). Such a difference can also be appreciated in Figure 7 (right), depicting the average number of traps $\langle N_t \rangle$ extracted from fitting the RTN distributions with a simplified model [32]: note that the departing of $\langle N_t \rangle$ from the initial value takes place at much higher cycling doses in the 3D NAND case than in the planar array. While this suggests an increased hardening of 3D cells against cycling-induced defects, it should not be forgotten that 3D cells feature a native trap density higher than their planar counterparts (see Figure 4, left), mostly due to the GB traps not present in crystalline silicon, which may hide the initial-stage growth of cycling-induced defects. It is also interesting to note that a stronger dependence on cycling in 3D arrays was instead reported in [119], which might be ascribed to either a larger trap generation rate due to different cycling conditions or to a lower number of native traps, as fewer traps in the NAND cells would result in a more noticeable increment due to cycling. Furthermore, a transient effect related to a non-stationary condition, as hinted by the asymmetric RTN distribution there reported (see [130] for discussion) could also affect the evaluation.

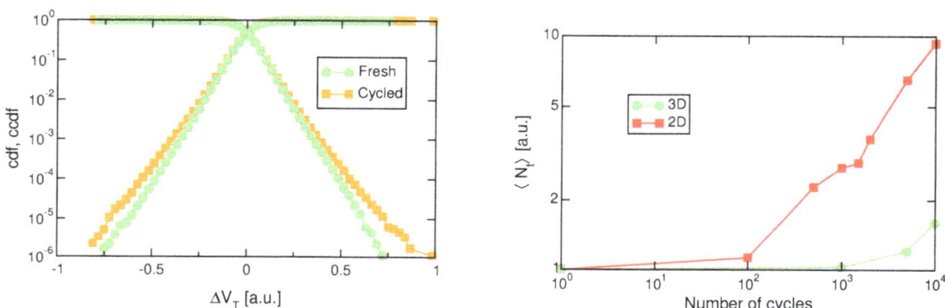

Figure 7. (**Left**) RTN cdf for a 3D NAND before and after cycling to 10k cycles [126], © 2018, IEEE. (**Right**) Average number of RTN traps as a function of cycling [126], © 2018, IEEE.

It is also interesting to note that data reported in Figure 7 were taken with programmed cells. However, in [110,131] a higher sensitivity of RTN to cycling was reported when cells are measured in the erased state (Figure 8). In the authors' view, this result is not a consequence of a different generation rate or annealing of stress-induced traps, but rather the result of different conduction profiles of the electrons as a consequence of the charge stored in the cells, enhancing the impact of newly-created traps at the interface. Such results demonstrate that the RTN picture is still not complete, notwithstanding the excellent work put forward by the scientific community.

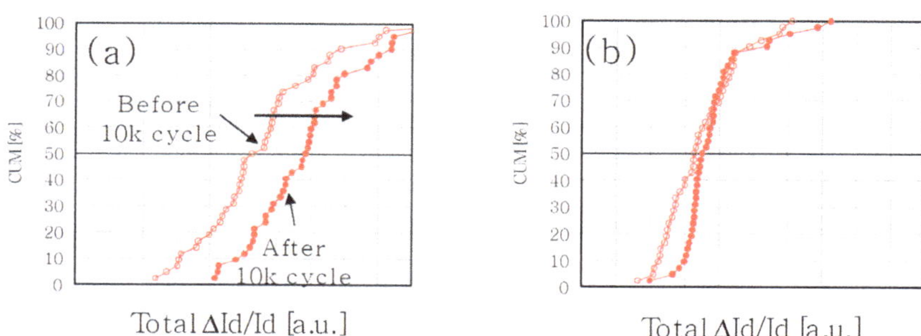

Figure 8. RTN cdf before and after cycling for the case of erased (**a**) and programmed (**b**) cells [110], © 2014, IEEE.

5. Conclusions

Ever since its first detection in MOS devices, RTN has retained two opposite faces, being a remarkable probe into the microscopic physics of carrier interactions with defects on one side, and a reliability threat on the other. It appears safe to say that even the transition to 3D NAND has not affected such characters, that are instead enhanced by the additional challenges built by the polycrystalline conduction channel. In this frame, this work has presented a review of the most significant experimental results in the field of random telegraph noise in 3D NAND, highlighting its current understanding and some open issues that require further efforts from the scientific and technological communities.

Funding: This research received no external funding.

Acknowledgments: The authors would like to acknowledge support from C. Miccoli, G. Paolucci, D. Resnati and G. Nicosia.

Conflicts of Interest: The authors declare no conflict of interest.

References

1. Ralls, K.S.; Skocpol, W.J.; Jackel, L.D.; Howard, R.E.; Fetter, L.A.; Epworth, R.W.; Tennant, D.M. Discrete resistance switching in submicrometer silicon inversion layers: Individual interface traps and low-frequency (1/f) noise. *Phys. Rev. Lett.* **1984**, *52*, 228–231. [CrossRef]
2. Uren, M.J.; Day, D.J.; Kirton, M.J. 1/f and random telegraph noise in silicon metal-oxide-semiconductor field-effect transistors. *Appl. Phys. Lett.* **1985**, *47*, 1195–1197. [CrossRef]
3. Howard, R.E.; Skocpol, W.J.; Jackel, L.D.; Mankiewich, P.M.; Fetter, L.A.; Tennant, D.M.; Epworth, R.; Ralls, K.S. Single electron switching events in nanometer-scale Si MOSFET's. *IEEE Trans. Electron Devices* **1985**, *32*, 1669–1674. [CrossRef]
4. Kirton, M.J.; Uren, M.J. Capture and emission kinetics of individual Si:SiO_2 interface states. *Appl. Phys. Lett.* **1986**, *48*, 1270–1272. [CrossRef]
5. Hung, K.K.; Ko, P.K.; Hu, C.; Cheng, Y.C. Flicker noise characterization of advanced MOS technologies. In Proceedings of the 1988 International Electron Devices Meeting, San Francisco, CA, USA, 11–14 December 1988; pp. 34–37.
6. Hung, K.K.; Ko, P.K.; Hu, C.; Cheng, Y.C. Random telegraph noise of deep-submicrometer MOSFET's. *IEEE Electron Device Lett.* **1990**, *11*, 90–92. [CrossRef]
7. Fang, P.; Hung, K.K.; Ko, P.K.; Hu, C. Hot-electron-induced traps studied through the random telegraph noise. *IEEE Electron Device Lett.* **1991**, *12*, 273–275. [CrossRef]
8. Simoen, E.; Dierickx, B.; Claeys, C.L.; Declerck, G.J. Explaining the amplitude of RTS noise in submicrometer MOSFET's. *IEEE Trans. Electron Devices* **1992**, *39*, 422–429. [CrossRef]
9. Tsai, M.H.; Ma, T.P.; Hook, T.B. Channel length dependence of random telegraph signal in sub-micron MOSFET's. *IEEE Electron Device Lett.* **1994**, *15*, 504–506. [CrossRef]
10. Uren, M.J.; Kirton, M.J.; Collins, S. Anomalous telegraph noise in small-area silicon metal-oxide-semiconductor field-effect transistors. *Phys. Rev. B* **1988**, *37*, 8346–8350. [CrossRef]
11. Kirton, M.J.; Uren, M.J. Noise in solid-state microstructures: A new perspective on individual defects, interface states and low-frequency (1/f) noise. *Adv. Phys.* **1989**, *38*, 367–468. [CrossRef]
12. Kirton, M.J.; Uren, M.J.; Collins, S.; Schultz, M.; Karmann, A.; Scheffer, K. Individual defects at the Si:SiO_2 interface. *Semicond. Sci. Technol.* **1989**, *4*, 1116–1126. [CrossRef]
13. Nakamura, H.; Yasuda, N.; Taniguchi, K.; Hamaguchi, C.; Toriumi, A. Existence of double-charged oxide traps in submicron MOSFET's. *Jpn. J. Appl. Phys.* **1989**, *28*, L2057–L2060. [CrossRef]
14. Ohata, A.; Toriumi, A.; Iwase, M.; Natori, K. Observation of random telegraph signals: Anomalous nature of defects at the Si/SiO_2 interface. *J. Appl. Phys.* **1990**, *68*, 200–204. [CrossRef]
15. Mueller, H.H.; Schulz, M. Conductance modulation of submicrometer metal-oxide-semiconductor field effect transistors by single-electron trapping. *J. Appl. Phys.* **1996**, *79*, 4178–4180. [CrossRef]
16. Mueller, H.H.; Schulz, M. Statistics of random telegraph noise in sub-μm MOSFETs. In Proceedings of the 14th International Conference: Noise in Physical Systems and 1/f Fluctuations, Leuven, Belgium, 14–18 July 1997; pp. 195–200.
17. Kurata, H.; Otsuga, K.; Kotabe, A.; Kajiyama, S.; Osabe, T.; Sasago, Y.; Narumi, S.; Tokami, K.; Kamohara, S.; Tsuchiya, O. The impact of random telegraph signals on the scaling of multilevel Flash memories. In Proceedings of the 2006 Symposium on VLSI Circuits, Honolulu, HI, USA, 15–17 June 2006; pp. 112–113.
18. Tega, N.; Miki, H.; Osabe, T.; Kotabe, A.; Otsuga, K.; Kurata, H.; Kamohara, S.; Tokami, K.; Ikeda, Y.; Yamada, R. Anomalously large threshold voltage fluctuation by complex random telegraph signal in floating gate Flash memory. In Proceedings of the 2006 International Electron Devices Meeting, San Francisco, CA, USA, 11–13 December 2006; pp. 491–494.

19. Gusmeroli, R.; Monzio Compagnoni, C.; Riva, A.; Spinelli, A.S.; Lacaita, A.L.; Bonanomi, M.; Visconti, A. Defects spectroscopy in SiO_2 by statistical random telegraph noise analysis. In Proceedings of the 2006 International Electron Devices Meeting, San Francisco, CA, USA, 11–13 December 2006; pp. 483–486.
20. Kurata, H.; Otsuga, K.; Kotabe, A.; Kajiyama, S.; Osabe, T.; Sasago, Y.; Narumi, S.; Tokami, K.; Kamohara, S.; Tsuchiya, O. Random telegraph signal in Flash memory: Its impact on scaling of multilevel Flash memory beyond the 90-nm node. *IEEE J. Solid-State Circuits* **2007**, *42*, 1362–1369. [CrossRef]
21. Fukuda, K.; Shimizu, Y.; Amemiya, K.; Kamoshida, M.; Hu, C. Random telegraph noise in Flash memories—Model and technology scaling. In Proceedings of the 2007 International Electron Devices Meeting, Washington, DC, USA, 10–12 December 2007; pp. 169–172.
22. Fantini, P.; Ghetti, A.; Marinoni, A.; Ghidini, G.; Visconti, A.; Marmiroli, A. Giant random telegraph signals in nanoscale floating-gate devices. *IEEE Electron Device Lett.* **2007**, *28*, 1114–1116. [CrossRef]
23. Miki, H.; Osabe, T.; Tega, N.; Kotabe, A.; Kurata, H.; Tokami, K.; Ikeda, Y.; Kamohara, S.; Yamada, R. Quantitative analysis of random telegraph signals as fluctuations of threshold voltages in scaled Flash memory cells. In Proceedings of the 2007 IEEE International Reliability Physics Symposium Proceedings—45th Annual Phoenix, AZ, USA, 15–19 April 2007; pp. 29–35.
24. Wong, H.S.; Taur, Y. Three-dimensional "atomistic" simulation of discrete random dopant distribution effects in sub-0.1 μm MOSFET's. In Proceedings of the IEEE International Electron Devices Meeting, Washington, DC, USA, 5–8 December 1993; pp. 705–708.
25. Asenov, A. Random dopant induced threshold voltage lowering and fluctuations in sub-0.1 μm MOSFET's: A 3-D "atomistic" simulation study. *IEEE Trans. Electron Devices* **1998**, *45*, 2505–2513. [CrossRef]
26. Asenov, A.; Brown, A.R.; Davies, J.H.; Saini, S. Hierarchical approach to "atomistic" 3-D MOSFET simulation. *IEEE Trans. Comput. Aided Des.* **1999**, *18*, 1558–1565. [CrossRef]
27. Asenov, A.; Saini, S. Suppression of random dopant-induced threshold voltage fluctuations in sub-0.1 μm MOSFET's with epitaxial and δ-doped channels. *IEEE Trans. Electron Devices* **1999**, *46*, 1718–1724. [CrossRef]
28. Vandamme, L.K.J.; Sodini, D.; Gingl, Z. On the anomalous behavior of the relative amplitude of RTS noise. *Solid-State Electron.* **1998**, *42*, 901–905. [CrossRef]
29. Asenov, A.; Balasubramaniam, R.; Brown, A.R.; Davies, J.H.; Saini, S. Random telegraph signal amplitudes in sub 100 nm (decanano) MOSFETs: A 3D "atomistic" simulation study. In Proceedings of the International Electron Devices Meeting, San Francisco, CA, USA, 10–13 December 2000; pp. 279–282.
30. Asenov, A.; Balasubramaniam, R.; Brown, A.R.; Davies, J.H. RTS amplitudes in decananometer MOSFETs: 3-D simulation study. *IEEE Trans. Electron Devices* **2003**, *50*, 839–845. [CrossRef]
31. Asenov, A.; Brown, A.R.; Davies, J.H.; Kaya, S.; Slavcheva, G. Simulation of intrinsic parameter fluctuations in decananometer and nanometer-scale MOSFETs. *IEEE Trans. Electron Devices* **2003**, *50*, 1837–1852. [CrossRef]
32. Monzio Compagnoni, C.; Gusmeroli, R.; Spinelli, A.S.; Lacaita, A.L.; Bonanomi, M.; Visconti, A. Statistical model for random telegraph noise in Flash memories. *IEEE Trans. Electron Devices* **2008**, *55*, 388–395. [CrossRef]
33. Ghetti, A.; Monzio Compagnoni, C.; Spinelli, A.S.; Visconti, A. Comprehensive analysis of random telegraph noise instability and its scaling in deca-nanometer Flash memories. *IEEE Trans. Electron Devices* **2009**, *56*, 1746–1752. [CrossRef]
34. Monzio Compagnoni, C.; Castellani, N.; Mauri, A.; Spinelli, A.S.; Lacaita, A.L. Three-dimensional electrostatics- and atomistic doping-induced variability of RTN time constants in nanoscale MOS devices—Part II: Spectroscopic implications. *IEEE Trans. Electron Devices* **2012**, *59*, 2495–2500. [CrossRef]
35. Castellani, N.; Monzio Compagnoni, C.; Mauri, A.; Spinelli, A.S.; Lacaita, A.L. Three-dimensional electrostatics- and atomistic doping-induced variability of RTN time constants in nanoscale MOS devices—Part I: Physical investigation. *IEEE Trans. Electron Devices* **2012**, *59*, 2488–2494. [CrossRef]
36. Adamu-Lema, F.; Monzio Compagnoni, C.; Amoroso, S.M.; Castellani, N.; Gerrer, L.; Markov, S.; Spinelli, A.S.; Lacaita, A.L.; Asenov, A. Accuracy and issues of the spectroscopic analysis of RTN traps in nanoscale MOSFETs. *IEEE Trans. Electron Devices* **2013**, *60*, 833–839. [CrossRef]
37. Spinelli, A.S.; Monzio Compagnoni, C.; Lacaita, A.L. Random telegraph noise in Flash memories. In *Noise in Nanoscale Semiconductor Devices*; Grasser, T., Ed.; Springer: Berlin, Germany, 2020; Chapter 6, pp. 201–227.
38. Monzio Compagnoni, C.; Goda, A.; Spinelli, A.S.; Feeley, P.; Lacaita, A.L.; Visconti, A. Reviewing the evolution of the NAND Flash technology. *Proc. IEEE* **2017**, *105*, 1609–1633. [CrossRef]
39. Tanaka, H.; Kido, M.; Yahashi, K.; Oomura, M.; Katsumata, R.; Kito, M.; Fukuzumi, Y.; Sato, M.; Nagata, Y.; Matsuoka, Y.; et al. Bit cost scalable technology with punch and plug process for ultra high density Flash memory. In Proceedings of the 2007 IEEE Symposium on VLSI Technology, Kyoto, Japan, 12–14 June 2007; pp. 14–15.
40. Fukuzumi, Y.; Katsumata, R.; Kito, M.; Kido, M.; Sato, M.; Tanaka, H.; Nagata, Y.; Matsuoka, Y.; Iwata, Y.; Aochi, H.; et al. Optimal integration and characteristics of vertical array devices for ultra-high density, bit-cost scalable Flash memory. In Proceedings of the International Electron Devices Meeting, San Francisco, CA, USA, 10–12 December 2007; pp. 449–452.
41. Maeda, T.; Itagaki, K.; Hishida, T.; Katsumata, R.; Kito, M.; Fukuzumi, Y.; Kido, M.; Tanaka, H.; Komori, Y.; Ishiduki, M.; et al. Multi-stacked 1 G cell/layer pipe-shaped BiCS Flash memory. In Proceedings of the 2009 Symposium on VLSI Circuits, Kyoto, Japan, 16–18 June 2009; pp. 22–23.

42. Ishiduki, M.; Fukuzumi, Y.; Katsumata, R.; Kito, M.; Kido, M.; Tanaka, H.; Komori, Y.; Nagata, Y.; Fujiwara, T.; Maeda, T.; et al. Optimal device structure for pipe-shaped BiCS Flash memory for ultra high density storage device with excellent performance and reliability. In Proceedings of the 2009 IEEE International Electron Devices Meeting, Baltimore, MD, USA, 7–9 December 2009; pp. 625–628.
43. Katsumata, R.; Kito, M.; Fukuzumi, Y.; Kido, M.; Tanaka, H.; Komori, Y.; Ishiduki, M.; Matsunami, J.; Fujiwara, T.; Nagata, Y.; et al. Pipe-shaped BiCS Flash memory with 16 stacked layers and multi-level-cell operation for ultra high density storage devices. In Proceedings of the 2009 Symposium on VLSI Technology, Kyoto, Japan, 15–17 June 2009; pp. 136–137.
44. Jang, J.; Kim, H.S.; Cho, W.; Cho, H.; Kim, J.; Shim, S.I.; Jang, Y.; Jeong, J.H.; Son, B.K.; Kim, D.W.; et al. Vertical cell array using TCAT (Terabit Cell Array Transistor) technology for ultra high density NAND Flash memory. In Proceedings of the 2009 Symposium on VLSI Technology, Kyoto, Japan, 15–17 June 2009; pp. 192–193.
45. Kim, W.; Choi, S.; Sung, J.; Lee, T.; Park, C.; Ko, H.; Jung, J.; Yoo, I.; Park, Y. Multi-layered vertical gate NAND Flash overcoming stacking limit for terabit density storage. In Proceedings of the 2009 Symposium on VLSI Technology, Kyoto, Japan, 15–17 June 2009; pp. 188–189.
46. Lue, H.T.; Chen, S.H.; Shih, Y.H.; Hsieh, K.Y.; Lu, C.Y. Overview of 3D NAND Flash and progress of vertical gate (VG) architecture. In Proceedings of the 2012 IEEE 11th International Conference on Solid-State and Integrated Circuit Technology, Xi'an, China, 29 October–1 November 2012; pp. 914–917.
47. Tanaka, T.; Helm, M.; Vali, T.; Ghodsi, R.; Kawai, K.; Park, J.K.; Yamada, S.; Pan, F.; Einaga, Y.; Ghalam, A.; et al. A 768Gb 3b/cell 3D-floating-gate NAND Flash memory. In Proceedings of the 2016 IEEE International Solid-State Circuits Conference, San Francisco, CA, USA, 31 January–4 February 2016; pp. 142–143.
48. Parat, K.; Goda, A. Scaling trends in NAND Flash. In Proceedings of the 2018 IEEE International Electron Devices Meeting, San Francisco, CA, USA, 1–5 December 2018; pp. 27–30.
49. Siau, C.; Kim, K.H.; Lee, S.; Isobe, K.; Shibata, N.; Verma, K.; Ariki, T.; Li, J.; Yuh, J.; Amarnath, A.; et al. A 512Gb 3-bit/cell 3D Flash memory on 128-wordline-layer with 132MB/s write performance featuring circuit-under-array technology. In Proceedings of the 2019 IEEE International Solid-State Circuits Conference, San Francisco, CA, USA, 17–21 February 2019; pp. 218–220.
50. Kim, D.H.; Kim, H.; Yun, S.; Song, Y.; Kim, J.; Joe, S.M.; Kang, K.H.; Jang, J.; Yoon, H.J.; Lee, K.; et al. A 1Tb 4 b/cell NAND Flash memory with t_{PROG} = 2 ms, t_R = 110 µs and 1.2 Gb/s high-speed IO rate. In Proceedings of the 2020 IEEE International Solid-State Circuits Conference, San Francisco, CA, USA, 16–20 February 2020.
51. Micheloni, R. (Ed.) *3D Flash Memories*; Springer: Berlin, Germany, 2016.
52. Spinelli, A.S.; Monzio Compagnoni, C.; Lacaita, A.L. Reliability of NAND Flash memories: Planar cells and emerging issues in 3D devices. *Computers* **2017**, *6*, 16. [CrossRef]
53. Monzio Compagnoni, C.; Spinelli, A.S. Reliability of NAND Flash arrays: A review of what the 2-D-to-3-D transition meant. *IEEE Trans. Electron Devices* **2019**, *66*, 4504–4516. [CrossRef]
54. Goda, A. 3D NAND technology achievements and future scaling perspectives. *IEEE Trans. Electron Devices* **2020**, *67*, 1373–1381. [CrossRef]
55. Whang, S.; Lee, K.; Shin, D.; Kim, B.; Kim, M.; Bin, J.; Han, J.; Kim, S.; Lee, B.; Jung, Y.; et al. Novel 3-Dimensional dual control-gate with surrounding floating-gate (DC-SF) NAND Flash cell for 1Tb file storage application. In Proceedings of the 2010 International Electron Devices Meeting, San Francisco, CA, USA, 6–8 December 2010; pp. 668–671.
56. Noh, Y.; Ahn, Y.; Yoo, H.; Han, B.; Chung, S.; Shim, K.; Lee, K.; Kwak, S.; Shin, S.; Choi, I.; et al. A new metal control gate last process (MCGL process) for high performance DC-SF (dual control gate with surrounding floating gate) 3D NAND Flash memory. In Proceedings of the 2012 Symposium on VLSI Technology (VLSIT), Honolulu, HI, USA, 12–14 June 2012; pp. 19–20.
57. Aritome, S.; Whang, S.; Lee, K.; Shin, D.; Kim, B.; Kim, M.; Bin, J.; Han, J.; Kim, S.; Lee, B.; et al. A novel three-dimensional dual control-gate with surrounding floating-gate (DC-SF) NAND flash cell. *Solid-State Electron.* **2013**, *79*, 166–171. [CrossRef]
58. Aritome, S.; Noh, Y.; Yoo, H.; Choi, E.S.; Joo, H.S.; Ahn, Y.; Han, B.; Chung, S.; Shim, K.; Lee, K.; et al. Advanced DC-SF cell technology for 3-D NAND Flash. *IEEE Trans. Electron Devices* **2013**, *60*, 1327–1333. [CrossRef]
59. Parat, K.; Dennison, C. A floating gate based 3-D NAND technology with CMOS under array. In Proceedings of the 2015 IEEE International Electron Devices Meeting, Washington, DC, USA, 7–9 December 2015; pp. 48–51.
60. Sako, M.; Watanabe, Y.; Nakajima, T.; Sato, J.; Muraoka, K.; Fujiu, M.; Kouno, F.; Nakagawa, M.; Masuda, M.; Kato, K.; et al. A low-power 64 Gb MLC NAND-Flash memory in 15 nm CMOS technology. *IEEE J. Solid-State Circuits* **2015**, *51*, 128–129.
61. Park, K.T.; Nam, S.; Kim, D.; Kwak, P.; Lee, D.; Choi, Y.H.; Choi, M.H.; Kwak, D.H.; Kim, D.H.; Kim, M.S.; et al. Three-dimensional 128 Gb MLC vertical NAND Flash memory with 24-WL stacked layers and 50 MB/s high-speed programming. *IEEE J. Solid-State Circuits* **2015**, *50*, 204–213. [CrossRef]
62. Kang, D.; Jeong, W.; Kim, C.; Kim, D.H.; Cho, Y.S.; Kang, K.T.; Ryu, J.; Kang, K.M.; Lee, S.; Kim, W.; et al. 256 Gb 3 b/cell V-NAND Flash memory with 48 stacked WL layers. *IEEE J. Solid-State Circuits* **2016**, *52*, 210–217. [CrossRef]
63. Lee, S.; Lee, J.; Park, I.; Park, J.; Yun, S.; Kim, M.; Lee, J.; Kim, M.; Lee, K.; Kim, T.; et al. A 128 Gb 2 b/cell NAND Flash memory in 14nm technology with t_{PROG} = 640 µs and 800 MB/s I/O rate. In Proceedings of the 2016 IEEE International Solid-State Circuits Conference, San Francisco, CA, USA, 31 January–4 February 2016; pp. 138–139.
64. Lee, S.; Kim, C.; Kim, M.; Joe, S.; Jang, J.; Kim, S.; Lee, K.; Kim, J.; Park, J.; Lee, H.J.; et al. A 1 Tb 4 b/cell 64-stacked-WL 3D NAND flash memory with 12 MB/s program throughput. In Proceedings of the 2018 IEEE International Solid-State Circuits Conference, San Francisco, CA, USA, 11–15 February 2018; pp. 340–342.

65. Seager, C.H. Grain boundaries in polycrystalline silicon. *Ann. Rev. Mater. Sci.* **1985**, *15*, 271–302. [CrossRef]
66. Jackson, W.B.; Johnson, N.M.; Biegelsen, D.K. Density of gap states of silicon grain boundaries determined by optical absorption. *Appl. Phys. Lett.* **1983**, *43*, 195–197. [CrossRef]
67. Werner, J.; Peisl, M. Exponential band tails in polycrystalline semiconductor films. *Phys. Rev. B* **1985**, *31*, 6681–6683. [CrossRef]
68. Fortunato, G.; Migliorato, P. Determination of gap state density in polycrystalline silicon by field-effect conductance. *Appl. Phys. Lett.* **1986**, *49*, 1025–1027. [CrossRef]
69. Evans, P.V.; Nelson, S.F. Determination of grain-boundary defect-state densities from transport measurements. *J. Appl. Phys.* **1991**, *69*, 3605–3611. [CrossRef]
70. Kimura, M.; Nozawa, R.; Inoue, S.; Shimoda, T.; Lui, B.O.K.; Tam, S.W.B.; Migliorato, P. Extraction of trap states at the oxide-silicon interface and grain boundary for polycrystalline silicon thin-film transistors. *Jpn. J. Appl. Phys.* **2001**, *40*, 5227–5236. [CrossRef]
71. Ikeda, H. Evaluation of grain boundary trap states in polycrystalline-silicon thin-film transistors by mobility and capacitance measurements. *J. Appl. Phys.* **2002**, *91*, 4637–4645. [CrossRef]
72. Hastas, N.A.; Tassis, D.H.; Dimitriadis, C.A.; Kamarinos, G. Determination of interface and bulk traps in the subthreshold region of polycrystalline silicon thin-film transistors. *IEEE Trans. Electron Devices* **2003**, *50*, 1991–1994. [CrossRef]
73. Kimura, M. Extraction of trap densities in entire bandgap of poly-Si thin-film transistors fabricated by solid-phase crystallization and dependence on process conditions of post annealing. *Solid-State Electron.* **2011**, *63*, 94–99. [CrossRef]
74. Wei, X.; Deng, W.; Fang, J.; Ma, X.; Huang, J. Determination of bulk and interface density of states in metal oxide semiconductor thin-film transistors by using capacitance–voltage characteristics. *Eur. Phys. J. Appl. Phys.* **2017**, *80*, 10103. [CrossRef]
75. Hack, M.; Shaw, J.G.; LeComber, P.G.; Willums, M. Numerical simulations of amorphous and polycrystalline silicon thin-film transistors. *Jpn. J. Appl. Phys.* **1990**, *29*, 2360–2362. [CrossRef]
76. Jacunski, M.D.; Shur, M.S.; Hack, M. Threshold voltage, field effect mobility, and gate-to-channel capacitance in polysilicon TFTs. *IEEE Trans. Electron Devices* **1996**, *43*, 1433–1440. [CrossRef]
77. Valdinoci, M.; Colalongo, L.; Baccarani, G.; Pecora, A.; Policicchio, I.; Fortunato, G.; Plais, F.; Legagneux, P.; Reita, C.; Priba, D. Analysis of electrical characteristics of polycrystalline silicon thin-film transistors under static and dynamic conditions. *Solid-State Electron.* **1997**, *41*, 1363–1369. [CrossRef]
78. Chow, T.; Wong, M. An analytical model for the transfer characteristics of a polycrystalline silicon thin-film transistor with a double exponential grain-boundary trap-state energy dispersion. *IEEE Electron Dev. Lett.* **2009**, *30*, 1072–1074. [CrossRef]
79. Amit, I.; Englander, D.; Horvitz, D.; Sasson, Y.; Rosenwaks, Y. Density and energy distribution of interface states in the grain boundaries of polysilicon nanowire. *Nano Lett.* **2014**, *14*, 6190–6194. [CrossRef]
80. Peisl, M.; Wieder, A.W. Conductivity in polycrystalline silicon—Physics and rigorous numerical treatment. *IEEE Trans. Electron Devices* **1983**, *30*, 1792–1797. [CrossRef]
81. Kim, D.M.; Khondker, A.; Ahmed, S.S.; Shah, R.R. Theory of conduction in polysilicon: Drift-diffusion approach in crystalline-amorphous-crystalline semiconductor system—Part I: Small signal theory. *IEEE Trans. Electron Devices* **1984**, *31*, 480–493. [CrossRef]
82. Khondker, A.N.; Kim, D.M.; Ahmed, S.S.; Shah, R.R. Theory of conduction in polysilicon: Drift-diffusion approach in crystalline-amorphous-crystalline semiconductor system—Part II: General I-V theory. *IEEE Trans. Electron Devices* **1984**, *31*, 493–500. [CrossRef]
83. Seto, J.Y.W. The electrical properties of polycrystalline silicon films. *J. Appl. Phys.* **1975**, *46*, 5247–5254. [CrossRef]
84. Baccarani, G.; Riccò, B.; Spadini, G. Transport properties of polycrystalline silicon films. *J. Appl. Phys.* **1978**, *49*, 5565–5570. [CrossRef]
85. Mandurah, M.M.; Saraswat, K.C.; Kamins, T.I. A model for conduction in polycrystalline silicon—Part I: Theory. *IEEE Trans. Electron Devices* **1981**, *28*, 1163–1171. [CrossRef]
86. Lu, N.C.C.; Gerzberg, L.; Lu, C.Y.; Meindl, J.D. Modeling and optimization of monolithic polycrystalline silicon resistors. *IEEE Trans. Electron Devices* **1981**, *28*, 818–830. [CrossRef]
87. Lee, J.Y.; Wang, F.Y. Temperature dependence of carrier trnsport in polycrystalline silicon. *Microelectron. J.* **1986**, *17*, 23–32. [CrossRef]
88. Mannara, A.; Spinelli, A.S.; Lacaita, A.L.; Monzio Compagnoni, C. Current transport in polysilicon-channel GAA MOSFETs: A modeling perspective. In Proceedings of the 49th European Solid-State Device Research Conference (ESSDERC), Cracow, Poland, 23–26 September 2019; pp. 222–225.
89. Mannara, A.; Malavena, G.; Spinelli, A.S.; Monzio Compagnoni, C. A comparison of modeling approaches for current transport in polysilicon-channel nanowire and macaroni GAA MOSFETs. *J. Comp. Electr.* **2021**, *20*, 537–544. [CrossRef]
90. Hsiao, Y.H.; Lue, H.T.; Chen, W.C.; Chen, C.P.; Chang, K.P.; Shih, Y.H.; Tsui, B.Y.; Lu, C.Y. Modeling the variability caused by random grain boundary and trap-location induced asymmetrical read behavior for a tight-pitch vertical gate 3D NAND Flash memory using double-gate thin-film transistor (TFT) device. In Proceedings of the 2012 International Electron Devices Meeting, San Francisco, CA, USA, 10–13 December 2012; pp. 609–612.
91. Hsiao, Y.H.; Lue, H.T.; Chen, W.C.; Chang, K.P.; Shih, Y.H.; Tsui, B.Y.; Hsieh, K.Y.; Lu, C.Y. Modeling the impact of random grain boundary traps on the electrical behavior of vertical gate 3-D NAND Flash memory devices. *IEEE Trans. Electron Devices* **2014**, *61*, 2064–2070. [CrossRef]

92. Yang, C.W.; Su, P. Simulation and investigation of random grain-boundary-induced variabilities for stackable NAND Flash using 3-D Voronoi grain patterns. *IEEE Trans. Electron Devices* **2014**, *61*, 1211–1214. [CrossRef]
93. Kim, J.; Lee, J.; Oh, H.; Rim, T.; Baek, C.K.; Meyyappan, M.; Lee, J.S. The variability due to random discrete dopant and grain boundary in 3D NAND unit cell. In Proceedings of the 2014 IEEE International Nanoelectronics Conference (INEC), Sapporo, Japan, 28–31 July 2014; pp. 66–68.
94. Wang, P.Y.; Tsui, B.Y. A novel approach using discrete grain-boundary traps to study the variability of 3-D vertical-gate NAND Flash memory cells. *IEEE Trans. Electron Devices* **2015**, *62*, 2488–2493. [CrossRef]
95. Degraeve, R.; Clima, S.; Putcha, V.; Kaczer, B.; Roussel, P.; Linten, D.; Groeseneken, G.; Arreghini, A.; Karner, M.; Kernstock, C.; et al. Statistical poly-Si grain boundary model with discrete charging defects and its 2D and 3D implementation for vertical 3D NAND channels. In Proceedings of the 2015 IEEE International Electron Devices Meeting, Washington, DC, USA, 7–9 December 2015; pp. 121–124.
96. Lun, Z.; Shen, L.; Cong, Y.; Du, G.; Liu, X.; Wang, Y. Investigation of the impact of grain boundary on threshold voltage of 3-D MLC NAND Flash memory. In Proceedings of the 2015 Silicon Nanoelectronics Workshop (SNW), Kyoto, Japan, 14–15 June 2015; pp. 35–37.
97. Oh, H.; Kim, J.; Lee, J.; Rim, T.; Baek, C.K.; Lee, J.S. Effects of single grain boundary and random interface traps on electrical variations of sub-30 nm polysilicon nanowire structures. *Microelectron. Eng.* **2016**, *149*, 113–116. [CrossRef]
98. Resnati, D.; Mannara, A.; Nicosia, G.; Paolucci, G.M.; Tessariol, P.; Lacaita, A.L.; Spinelli, A.S.; Monzio Compagnoni, C. Temperature activation of the string current and its variability in 3-D NAND Flash arrays. In Proceedings of the 2017 IEEE International Electron Devices Meeting, San Francisco, CA, USA, 2–6 December 2017; pp. 103–106.
99. Nicosia, G.; Mannara, A.; Resnati, D.; Paolucci, G.M.; Tessariol, P.; Lacaita, A.L.; Spinelli, A.S.; Goda, A.; Monzio Compagnoni, C. Inpact of temperature on the amplitude of RTN fluctuations in 3-D NAND Flash cells. In Proceedings of the 2017 IEEE International Electron Devices Meeting, San Francisco, CA, USA, 2–6 December 2017; pp. 521–524.
100. Resnati, D.; Mannara, A.; Nicosia, G.; Paolucci, G.M.; Tessariol, P.; Spinelli, A.S.; Lacaita, A.L.; Monzio Compagnoni, C. Characterization and modeling of temperature effects in 3-D NAND Flash arrays—Part I: Polysilicon-induced variability. *IEEE Trans. Electron Devices* **2018**, *65*, 3199–3206. [CrossRef]
101. Nicosia, G.; Mannara, A.; Resnati, D.; Paolucci, G.M.; Tessariol, P.; Spinelli, A.S.; Lacaita, A.L.; Goda, A.; Monzio Compagnoni, C. Characterization and modeling of temperature effects in 3-D NAND Flash arrays—Part II: Random telegraph noise. *IEEE Trans. Electron Devices* **2018**, *65*, 3207–3213. [CrossRef]
102. Verreck, D.; Arreghini, A.; Schanovsky, F.; Stanojević, Z.; Steiner, K.; Mitterbauer, F.; Karner, M.; den Bosch, G.V.; Furnémont, A. 3D TCAD model for poly-Si channel current and variability in vertical NAND Flash memory. In Proceedings of the 2019 International Conference on Simulation of Semiconductor Processes and Devices (SISPAD), Udine, Italy, 4–6 September 2019; pp. 61–64.
103. Degraeve, R.; Toledano-Luque, M.; Arreghini, A.; Tang, B.; Capogreco, E.; Lisoni, J.; Roussel, P.; Kaczer, B.; Van den bosch, G.; Groeseneken, G.; et al. Characterizing grain size and defect energy distribution in vertical SONOS poly-Si channels by means of a resistive network model. In Proceedings of the 2013 IEEE International Electron Devices Meeting, Washington, DC, USA, 9–11 December 2013; pp. 558–561.
104. Yoo, W.S.; Ishigaki, T.; Ueda, T.; Kang, K.; Kwak, N.Y.; Sheen, D.S.; Kim, S.S.; Ko, M.S.; Shin, W.S.; Lee, B.S.; et al. Grain size monitoring of 3D Flash memory channel poly-Si using multiwavelength Raman spectroscopy. In Proceedings of the 2014 14th Annual Non-Volatile Memory Technology Symposium (NVMTS), Jeju, Korea, 27–29 October 2014; pp. 44–47.
105. Kim, S.Y.; Park, J.K.; Hwang, W.S.; Lee, S.J.; Lee, K.H.; Pyi, S.H.; Cho, B.J. Dependence of grain size on the performance of a polysilicon channel TFT for 3D NAND Flash memory. *J. Nanosci. Nanotechnol.* **2016**, *16*, 5044–5048. [CrossRef] [PubMed]
106. He, D.; Okada, N.; Fortmann, C.M.; Shimizu, I. Carrier transport in polycrystalline silicon films deposited by a layer-by-layer technique. *J. Appl. Phys.* **1994**, *76*, 4728–4733. [CrossRef]
107. Walker, A.J.; Herner, S.B.; Kumar, T.; Chen, E.H. On the conduction mechanism in polycrystalline silicon thin-film transistors. *IEEE Trans. Electron Devices* **2004**, *51*, 1856–1866. [CrossRef]
108. Jeong, M.K.; Joe, S.M.; Seo, C.S.; Han, K.R.; Choi, E.; Park, S.K.; Lee, J.H. Analysis of random telegraph noise and low frequency noise properties in 3-D stacked NAND Flash memory with tube-type poly-Si channel structure. In Proceedings of the 2012 Symposium on VLSI Technology (VLSIT), Honolulu, HI, USA, 12–14 June 2012; pp. 55–56.
109. Jeong, M.K.; Joe, S.M.; Jo, B.S.; Kang, H.J.; Bae, J.H.; Han, K.R.; Choi, E.; Cho, G.; Park, S.K.; Park, B.G.; et al. Characterization of traps in 3-D stacked NAND Flash memory devices with tube-type poly-Si channel structure. In Proceedings of the 2012 International Electron Devices Meeting, San Francisco, CA, USA, 10–13 December 2012.
110. Kang, D.; Lee, C.; Hur, S.; Song, D.; Choi, J.H. A new approach for trap analysis of vertical NAND Flash cell using RTN characteristics. In Proceedings of the 2014 IEEE International Electron Devices Meeting, San Francisco, CA, USA, 15–17 December 2014; pp. 367–370.
111. Amoroso, S.M.; Monzio Compagnoni, C.; Ghetti, A.; Gerrer, L.; Spinelli, A.S.; Lacaita, A.L.; Asenov, A. Investigation of the RTN distribution of nanoscale MOS devices from subthreshold to on-state. *IEEE Electron Device Lett.* **2013**, *34*, 683–685. [CrossRef]
112. Saraza-Canflanca, P.; Martin-Martinez, J.; Castro-Lopez, R.; Roca, E.; Rodriguez, R.; Nafria, M.; Fernandez, F.V. A detailed study of the gate/drain voltage dependence of RTN in bulk pMOS transistors. *Microelectron. Eng.* **2019**, *215*, 111004. [CrossRef]

113. Spinelli, A.S.; Monzio Compagnoni, C.; Lacaita, A.L. Variability effects in nanowire and macaroni MOSFETs—Part II: Random telegraph noise. *IEEE Trans. Electron Devices* **2020**, *67*, 1492–1497. [CrossRef]
114. Andrade, M.; Toledano-Luque, M.; Fourati, F.; Degraeve, R.; Martino, J.; Claeys, C.; Simoen, E.; Van den bosch, G.; Van Houdt, J. RTN assessment of traps in polysilicon cylindrical vertical FETs for NVM application. *Microelectron. Eng.* **2013**, *109*, 105–108. [CrossRef]
115. Toledano-Luque, M.; Degraeve, R.; Roussel, P.J.; Luong, V.; Tang, B.; Lisoni, J.G.; Tan, C.L.; Arreghini, A.; Van den bosch, G.; Groeseneken, G.; et al. Statistical spectroscopy of switching traps in deeply scaled vertical poly-Si channel for 3D memories. In Proceedings of the 2013 IEEE International Electron Devices Meeting, Washington, DC, USA, 9–11 December 2013; pp. 562–565.
116. Lee, C.M.; Tsui, B.Y. Random telegraph signal noise arising from grain boundary traps in nano-scale poly-Si nanowire thin-film transistors. In Proceedings of the 2013 International Symposium on VLSI Technology, Systems and Application (VLSI-TSA), Hsinchu, Taiwan, 22–24 April 2013; pp. 47–48.
117. Nicosia, G.; Goda, A.; Spinelli, A.S.; Monzio Compagnoni, C. Investigation of the temperature dependence of random telegraph noise fluctuations in nanoscale polysilicon-channel 3-D Flash cells. *Solid-State Electron.* **2019**, *151*, 18–22. [CrossRef]
118. Degraeve, R.; Toledano-Luque, M.; Suhane, A.; Van den bosch, G.; Arreghini, A.; Tang, B.; Kaczer, B.; Roussel, P.; Kar, G.S.; Van Houdt, J.; et al. Statistical characterization of current paths in narrow poly-Si channels. In Proceedings of the 2011 International Electron Devices Meeting, Washington, DC, USA, 5–7 December 2011; pp. 287–290.
119. Nowak, E.; Kim, J.H.; Kwon, H.; Kim, Y.G.; Sim, J.S.; Lim, S.H.; Kim, D.S.; Lee, K.H.; Park, Y.K.; Choi, J.H.; et al. Intrinsic fluctuations in vertical NAND Flash memories. In Proceedings of the 2012 Symposium on VLSI Technology (VLSIT), Honolulu, HI, USA, 12–14 June 2012; pp. 21–22.
120. Toledano-Luque, M.; Degraeve, R.; Kaczer, B.; Tang, B.; Roussel, P.J.; Weckx, P.; Franco, J.; Arreghini, A.; Suhane, A.; Kar, G.S.; et al. Quantitative and predictive model of reading current variability in deeply scaled vertical poly-Si channel for 3D memories. In Proceedings of the 2012 International Electron Devices Meeting, San Francisco, CA, USA, 10–13 December 2012.
121. Goda, A.; Miccoli, C.; Monzio Compagnoni, C. Time dependent threshold-voltage fluctuations in NAND Flash memories: From basic physics to impact on array operation. In Proceedings of the 2015 IEEE International Electron Devices Meeting, Washington, DC, USA, 7–9 December 2015; pp. 374–377.
122. Chou, Y.L.; Wang, T.; Lin, M.; Chang, Y.W.; Liu, L.; Huang, S.W.; Tsai, W.J.; Lu, T.C.; Chen, K.C.; Lu, C.Y. Poly-silicon trap position and pass voltage effects on RTN amplitude in a vertical NAND Flash cell string. *IEEE Electron Dev. Lett.* **2016**, *37*, 998–1001. [CrossRef]
123. Hsieh, C.C.; Lue, H.T.; Hsu, T.H.; Du, P.Y.; Chiang, K.H.; Lu, C.Y. A Monte Carlo simulation method to predict large-density NAND product memory window from small-array test element group (TEG) verified on a 3D NAND Flash test chip. In Proceedings of the 2016 IEEE Symposium on VLSI Technology, Honolulu, HI, USA, 14–16 June 2016; pp. 63–64.
124. Hung, C.H.; Lue, H.T.; Hung, S.N.; Hsieh, C.C.; Chang, K.P.; Chen, T.W.; Huang, S.L.; Chen, T.S.; Chang, C.S.; Yeh, W.W.; et al. Design innovations to optimize the 3D stackable vertical gate (VG) NAND Flash. In Proceedings of the 2012 International Electron Devices Meeting, San Francisco, CA, USA, 10–13 December 2012; pp. 227–230.
125. Resnati, D.; Goda, A.; Nicosia, G.; Miccoli, C.; Spinelli, A.S.; Monzio Compagnoni, C. Temperature effects in NAND Flash memories: A comparison between 2-D and 3-D arrays. *IEEE Electron Dev. Lett.* **2017**, *38*, 461–464. [CrossRef]
126. Nicosia, G.; Goda, A.; Spinelli, A.S.; Monzio Compagnoni, C. Impact of cycling on random telegraph noise in 3-D NAND Flash arrays. *IEEE Electron Dev. Lett.* **2018**, *39*, 1175–1178. [CrossRef]
127. Cappelletti, P. Non volatile memory evolution and revolution. In Proceedings of the 2015 IEEE International Electron Devices, Washington, DC, USA, 7–9 December 2015; pp. 241–244.
128. Miccoli, C.; Paolucci, G.M.; Monzio Compagnoni, C.; Spinelli, A.S.; Goda, A. Cycling pattern and read/bake conditions dependence of random telegraph noise in decananometer NAND Flash arrays. In Proceedings of the 2015 IEEE International Reliability Physics Symposium, Monterey, CA, USA, 19–23 April 2015; pp. MY.9.1–MY.9.4.
129. Monzio Compagnoni, C.; Spinelli, A.S.; Beltrami, S.; Bonanomi, M.; Visconti, A. Cycling effect on the random telegraph noise instabilities of NOR and NAND Flash arrays. *IEEE Electron Device Lett.* **2008**, *29*, 941–943. [CrossRef]
130. Monzio Compagnoni, C.; Gusmeroli, R.; Spinelli, A.S.; Lacaita, A.L.; Bonanomi, M.; Visconti, A. Statistical investigation of random telegraph noise I_D instabilities in Flash cells at different initial trap-filling conditions. In Proceedings of the 2007 45th Annual IEEE International Reliability Physics Symposium Proceedings, Phoenix, AZ, USA, 15–19 April 2007; pp. 161–166.
131. Choi, N.; Kang, H.J.; Lee, J.H. Analysis of random telegraph noise characteristics with two different cell states in 3-D NAND Flash memory. *J. Semicond. Technol. Sci.* **2019**, *19*, 153–157. [CrossRef]

Article

An SVM-Based NAND Flash Endurance Prediction Method

Haichun Zhang [1], Jie Wang [2], Zhuo Chen [1], Yuqian Pan [3], Zhaojun Lu [3] and Zhenglin Liu [1,*]

1 School of Optical and Electronic Information, Huazhong University of Science and Technology, Wuhan 430074, China; D201880640@hust.edu.cn (H.Z.); zoarzzz@outlook.com (Z.C.)
2 Shenzhen Kaiyuan Internet Security Technology Co., Ltd., Shenzhen 518000, China; wangjie@seczone.cn
3 School of Cyber Science and Engineering, Huazhong University of Science and Technology, Wuhan 430074, China; yuqianpan@126.com (Y.P.); lzj_cse@hust.edu.cn (Z.L.)
* Correspondence: liuzhenglin@hust.edu.cn

Abstract: NAND flash memory is widely used in communications, commercial servers, and cloud storage devices with a series of advantages such as high density, low cost, high speed, anti-magnetic, and anti-vibration. However, the reliability is increasingly getting worse while process improvements and technological advancements have brought higher storage densities to NAND flash memory. The degradation of reliability not only reduces the lifetime of the NAND flash memory but also causes the devices to be replaced prematurely based on the nominal value far below the minimum actual value, resulting in a great waste of lifetime. Using machine learning algorithms to accurately predict endurance levels can optimize wear-leveling strategies and warn bad memory blocks, which is of great significance for effectively extending the lifetime of NAND flash memory devices and avoiding serious losses caused by sudden failures. In this work, a multi-class endurance prediction scheme based on the SVM algorithm is proposed, which can predict the remaining P-E cycle level and the raw bit error level after various P-E cycles. Feature analysis based on endurance data is used to determine the basic elements of the model. Based on the error features, we present a variety of targeted optimization strategies, such as extracting the numerical features closely related to the endurance, and reducing the noise interference of transient faults through short-term repeated operations. Besides a high-parallel flash test platform supporting multiple protocols, a feature preprocessing module is constructed based on the ZYNQ-7030 chip. The pipelined module of SVM decision model can complete a single prediction within 37 us.

Keywords: NAND flash memory; test platform; endurance; support vector machine; raw bit error

Citation: Zhang, H.; Wang, J.; Chen, Z.; Pan, Y.; Lu, Z.; Liu, Z. An SVM-Based NAND Flash Endurance Prediction Method. *Micromachines* **2021**, *12*, 746. https://doi.org/10.3390/mi12070746

Academic Editors: Cristian Zambelli and Rino Micheloni

Received: 21 May 2021
Accepted: 21 June 2021
Published: 25 June 2021

Publisher's Note: MDPI stays neutral with regard to jurisdictional claims in published maps and institutional affiliations.

1. Introduction

With the development of smart devices and cloud computing, flash memory has gained great popularity in various fields [1]. NAND flash memory has achieved larger storage capacity and higher storage speed than NOR flash memory by virtue of the design mode of storage units connected in series, becoming an important large-scale data storage medium. In order to pursue higher storage density, a variety of technologies have been developed in the field of NAND flash memory. Three-dimensional structure technology [2] is committed to transforming a planar structure into a three-dimensional structure, which increases the storage capacity under the same area. Multi-bit memory cell technology [3] focuses on improving the number of bits in the storage unit in order to achieve a multiple increase in storage capacity. With gradual in-depth study of the two technologies, researchers have found that while the storage density of NAND flash memory has doubled, the data reliability problem has worsened.

Data reliability marks the accuracy of data storage. If data errors occur during use, serious consequences will be immeasurable. In the field of NAND flash memory, data reliability problems are mainly reflected in retention [4] and endurance. The former reflects the data retention time without re-erasing, while the latter is the problem of reliability degradation caused by structural damage to the memory cell during use. Compared to retention, endurance has a greater impact on the actual product. As the number of programming increases, the endurance of the flash memory decreases and the number Raw Bit Error (RBE) gradually rises. The RBE number is the number of bits of difference between the actual read data and the actual programmed data without error correction, which is an important parameter to characterize the degree of endurance change. When the RBE numbers exceeds a certain limit, the flash memory will not continue to be used normally.

It has become an important direction for academia to suppress data errors caused by reduced endurance. The Error Correcting Code (ECC) error correction algorithm uses special coding rules to check and correct the original data [5], so the endurance of the flash memory can be indirectly improved by optimizing the error correction algorithm. However, limited by the storage space, the optimization effect is not ideal. Wear-Leveling technology [6] indirectly delays the occurrence of user data failure by balancing the number of programming and erasing of each block. However, the total endurance of the regions in the flash memory is not equal; there is room for further optimization of the technology. The Read-Retry technology reduces data errors by modifying the hard-decision reference voltage of the read operation, but this will increase the operating time and reduce the performance.

However, the real dilemma of flash memory reliability research lies in the uncertainty of endurance, which will lead to huge waste. The minimum actual endurance of flash memory is often dozens or even a hundred times the manufacturer's nominal value [7]. The reason why manufacturers specify the nominal value so conservatively is the huge difference in endurance between the same type of flash memory particles [8]. Even if the manufacturer inspects and screens the wafers before they leave the factory, there are still several times or even ten times the endurance difference between the same model and the same batch of flash memory particles. Besides, the particle-level inspection is destructive and extremely time-consuming. If the endurance can be accurately predicted and an early warning be made, the user can adjust the critical value of the data transfer process and greatly extend the service life of the flash memory.

Using machine learning algorithms to accurately predict changes in flash memory endurance-related parameters have become an important means to solve the flash memory reliability dilemma. It can greatly optimize the existing flash memory management strategies and implement accurate endurance warnings. Damien Hogan tried to combine a supervised Genetic Programming (GP) algorithm with the endurance prediction of 2D Multi-level Cell (MLC) flash memory to determine whether the sample flash memory with different levels of Program-Erase (P-E) cycles will generate uncorrectable data errors [9]. However, the GP two-category prediction model finally obtained in the study has a prediction accuracy of only 83.5% on the test set when the decision boundary is 35,000. Barry Fitzgerald observed through a large amount of experimental data that the word line (WL) number, page type, and page parity in MLC flash memory will affect the code word error rate (CWER), the programming, and erasing duration [10]. Using the feature, the study proposed a sampling method based on the error probability density function [11], and constructed eight different two-class machine learning models. However, the study neglected the class balance of the data set. The number of negative samples representing the number of codeword errors exceeding 100 only accounted for 0.03% of the total number of samples, which led to a significant decrease in the reliability of the model accuracy results. Ruixiang Ma considered that the predictive model may lose its validity due to changes in the flash memory usage environment, so the incremental changes in endurance parameters are used to update the predictive model to adapt to the parameter changes at different endurance stages [12]. However, this solution did not take into account the hardware complexity and application limitations of using the same flash memory pre-data to predict the later endurance.

On the one hand, the endurance prediction model established by existing studies performs two-class prediction of the RBE numbers, and the RBE numbers corresponding to the classification boundary is close to the upper limit of the ECC error correction algorithm, which limits the application scenarios of the prediction model. On the other hand, existing research does not consider the disturbance of electrical effects such as transient errors in NAND flash memory on the prediction results, which greatly reduces the prediction accuracy. In order to solve the endurance prediction problem, this paper designs a set of NAND flash memory endurance class prediction method based on SVM algorithm based on a large amount of experimental test data and combined with micro-mechanism analysis. The main contents of the paper are as follows:

(1) The multi-classification model describes the endurance level with the remaining P-E cycle number and RBE numbers. The output result has multiple levels and labels, which greatly expands the application scenarios.
(2) This paper has carried out a number of targeted optimization strategies for the disturbance factors in the flash memory. By preprocessing the input features, the prediction model still has a high accuracy rate when the output categories increase.
(3) We further use FPGA to implement the SVM prediction circuit module, which can be actually applied to solid-state hard disk controller chips. Compared with common embedded software implementations, the prediction circuit module designed in the paper has a faster prediction speed, with less hardware resources. and has broad application prospects.

2. Flash Error Mechanism and Prediction Algorithm

2.1. Flash Error Mechanism

The cause of the NAND flash memory endurance problem lies in its unique memory cell and array structure. The basic memory cell structure of flash memory is based on the development and evolution of the Floating Gate (FG) Field Effect Transistor (FET). The internal electrical disturbance phenomenon of the NAND flash memory is closely related to its array structure. The NAND flash memory will produce the decrease of reliability due to many physical effects in actual scenarios. There are several kinds of mechanisms.

2.1.1. Unit Wear-Out

Unit cell wear is the direct cause of reduced endurance of flash memory. Wear-out and aging often occur in the process of charge tunneling and transfer during programming and erasing operations. Wear-out causes the atomic bonds at the interface between the charge trap layer and the insulating layer to break, resulting in interface traps that interfere with the charge transport process and cause the threshold voltage to deviate from the ideal value. Therefore, the interface traps are the main reason for the endurance of the flash memory cell to decrease and the occurrence of error bit flips. The wear-out and aging of the unit caused by the P-E cycle is small but irreversible, and the number of P-E cycles is correlated with the endurance.

2.1.2. Disturbance

Certain electrical effects caused by the special array structure of flash memory can cause threshold voltage shifts. The most common effect is the disturbance phenomenon [13]. Disturbance is not permanently structurally destructive. The memory cell is restored to its original state by erasing. And the severity of disturbance is closely related to the programming pattern. The specific programming pattern significantly stimulates some disturbances. Multiple read operations on the same memory cell before the erase operation will cause reads disturbance, which causes the threshold voltage of the affected memory cell to shift in the positive direction. When the shifted threshold voltage exceeds the hard-decision reference voltage between different programming states, data errors will occur. Program disturbance and pass disturbance will occur during the programming operation. Edge word line disturbance also occurs during programming operations. The edge word line unit generates a large number of electron-hole pairs due to the large gate-induced drain leakage [14]. The electrons are accelerated to the channel and injected into the storage layer in the edge word line unit, leading to a surge in threshold voltage and data errors.

2.1.3. Transient Error

Transient error refers to the data error flipping caused by some uncertain transient factors during the operation of the flash memory. Because of the uncertainty of inducing factors, these transient errors are difficult to limit by conventional means. The most typical transient error is the uncertainty error caused by the Random Telegraph Noise (RTN) phenomenon. The error causes uncertain fluctuations in the drain current [15], which in turn causes the threshold voltage to fluctuate uncertainly.

2.2. Prediction Algorithm

The SVM algorithm has been developed rapidly through the efforts of many scholars since it was proposed in 1963. It is based on slack variables [16] and VC dimension [17], and has strong sparsity and generalization capabilities. The ultimate goal of linear SVM is to find the optimal hyperplane $P_k : \omega^T X + b = 0$ that can divide the sample space correctly. However, the actual classification problem is often a non-linear problem, which requires a certain non-linear transformation to achieve spatial upscaling, so that a hyperplane that can be correctly classified in the high-dimensional feature space reappears. The kernel function can reduce the complexity of high-dimensional inner product operations. The SVM algorithm often uses Radial Basis Function as a kernel function in practical applications.

The SVM decision model only supports two classifications. The One-against-One (OAO) method [18] groups the training data according to the output categories, and builds a separate two-class SVM model between each two categories. A total of $K(K-1)/2$ decision boundaries are obtained, and K is the total number of categories. When facing the new data to be classified, the OAO method inputs it into $K(K-1)/2$ models to obtain the corresponding classification results, respectively. Finally, according to the voting strategy, the category with the most votes among all the results is counted as the final classification results.

The Decision Tree (DT) classifier can perform fast and effective classification in the face of a large amount of data input. However, weak generalization will cause serious overfitting in the sample space where the total number of categories is unbalanced. At the same time, the instability of the DT classifier will cause it to be very sensitive to high-frequency jitter in the flash memory training data, and generate very different DT models.

The implementation method, the K-Nearest Neighbor (KNN) classifier, is simple, and for the simplest KNN classifier, it does not even need to be trained. But if the training set sample points are not clipped, the method needs to store all sample points and calculate the distance from the sample points to be classified to all the sample points. The storage space and calculation resources required are very huge, which is not suitable for circuit realization.

3. Endurance Prediction Method

3.1. The Process of Endurance Prediction Method

Figure 1 shows the process of the SVM-based endurance prediction method; the SVM-based NAND Flash endurance prediction method includes two phases: Training phase and testing phase. The purpose of the training phase is to obtain sample data sets and use machine learning algorithms to establish a decision model suitable for Flash endurance prediction. The testing phase uses the established decision model to predict endurance.

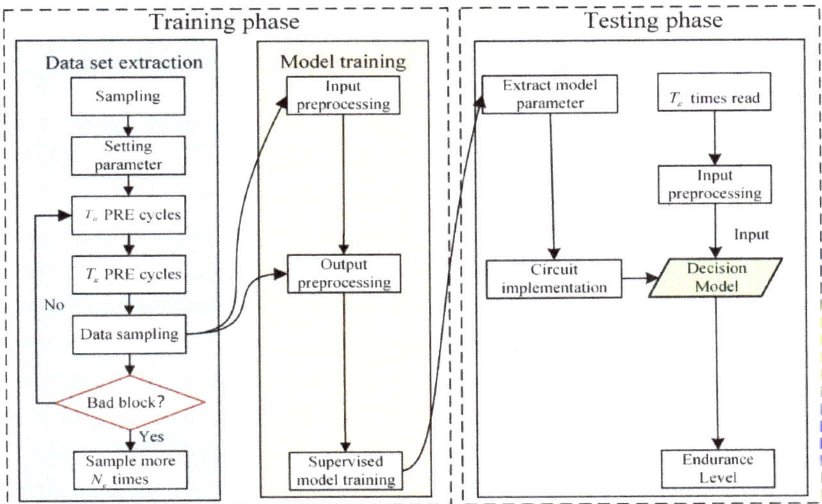

Figure 1. The process of SVM-based endurance prediction method.

3.1.1. Training Phase

Training phase includes data set extraction and model training. Data set extraction performs repeated P-E cycles on memory blocks of different flash memory particles with the same model to obtain flash memory endurance-related data in a certain rule. Model training uses the acquired flash memory endurance-related data set to train the machine learning model to obtain the decision function.

(a) Sample Selection

Select a certain number of flash memory blocks of the same model with suitable locations as samples. In order to avoid over-fitting, the number of flash memory blocks selected in each flash memory particle needs to be consistent.

(b) Parameter Setting

Set flash memory specification information, including interface protocol type, storage unit type, block size, page size, and the total number of blocks in a single logical unit. At the same time, set the test information, such as the used programming pattern, test mode, bad block characteristic error rate, etc. The random programming pattern can simulate various programming levels and combinations, fitting the programming pressure in practical applications to the greatest extent, which sets it as the default programming pattern in the proposed prediction method.

(c) P-E Cycle

T_α P-E cycles are performed on flash memory particles to be tested to accelerate the wear of the memory unit. The number of P-E cycles N_c will be increased by one each time a cycle is completed. When performing P-E cycles on flash memory particles to accelerate memory cell wear, it is necessary to keep the idle period interval between programming and erasing operations in each cycle fixed to eliminate the difference in actual endurance caused by different idle period intervals.

(d) P-R-E Cycle and Data Sampling

T_ε Program-Read-Erase (P-R-E) cycles are performed on the flash memory particles to be tested to obtain the data required for modeling. Update the current cycle number $N_c = N_c' + T_\varepsilon$, and N_c' is the cycle number before the update. Multiple P-R-E cycle operations in a short period of time help reduce the negative effects of transient errors. Each

P-R-E cycle compares the read data with the written original data to obtain the RBE numbers of each flash page. At the same time, the current cycle number N_c and the duration of the flash operation are recorded during each P-R-E cycle as the original model training data set.

Repeat steps c and d until it is detected that the RBE numbers of a page in the block exceeds the ECC error correction. In order to ensure that there are enough samples in each endurance stage, continue to perform T_e PE cycles and then stop sampling.

3.1.2. Testing Phase

The main purpose of the actual application process of the model is to extract the parameters of the trained model and implement it with a specific circuit, and then face the new data in the actual use scene, call the prediction model circuit, and get the prediction result.

3.2. Analysis of Prediction Model

3.2.1. Prediction Object Selection

The object of programming operation is flash memory page, and scholars have focused more on prediction research in the past. However, various disturbances seriously affect the accuracy of these prediction models in actual scenarios. This paper takes the flash memory block as an independent prediction object. The decision was based on the following reasons:

1. Obvious Endurance Difference between Flash Memory Pages

As shown in Figure 2a, take Intel's 29F32B2ALCMG3 NAND Flash particles as an example, different pages show different RBE numbers. In the interval where the page number is lower than 200, the RBE numbers of some pages is significantly higher, showing a more obvious trend of rising with the increase in number of P-E cycles. While other flash memory pages have significantly lower RBE numbers, the change trend is also very irregular. Even after ignoring the high-frequency jitter, there is a local feature where the RBE numbers decreases with the increase in the number of P-E cycles.

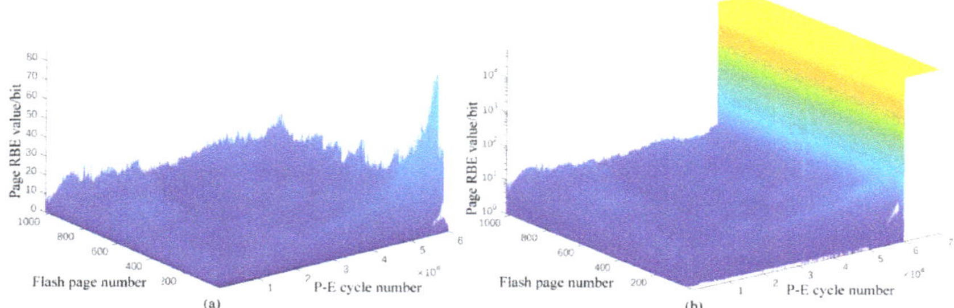

Figure 2. The relationship between the RBE on the page and the P-E cycle number: (**a**) Early and mid in lifecycle; (**b**) End of lifecycle.

The memory cells of different pages have differences in structural size and attributes [19], which leads to inconsistency of the actual tunneling charges suffered by different pages under the same macro programming pressure, directly affecting the degree of cell wear. Moreover, the new three-dimensional multilayer process will bring more serious physical structure differences and disturbance effects [20], resulting in more significant differences in the endurance of flash memory pages. If the flash memory page is used as the prediction object for modeling, the difference in the degree of wear and change trend will greatly reduce the accuracy of the model's prediction results.

2. The Integrity of the Flash Memory Block

There is also a significant integrity between different pages in the same flash memory block, such as the "cliff" phenomenon. As shown in Figure 2b, even if the wear degree and RBE numbers of different pages are different, the RBE numbers of all pages in the flash memory block jumps at the same time at the end of the life and surges to more than 60,000. Considering that the page capacity of the selected flash is 16 KB and the pattern used is pseudo-random, page RBE numbers as high as half of the page capacity means that the block has lost its storage function. The structural traps generated by the cell wear form fine "cracks" [21], which is very common in many types of flash memory particles. When the "cracks" accumulate to a certain extent, the insulating layer is broken down, forming a penetration path, which causes a large area of memory cells to fail. Pages with large differences in physical characteristics show the same end-of-life endurance performance, which makes it difficult to predict the endurance of the flash memory page as an independent object.

3. Array Coupling and Bad Block Management

On the one hand, the word line wear in a multi-bit memory cell will be reflected in the RBE number of multiple pages, which means that coupling relationship between different pages affects each other. On the other hand, the difference and randomness of the written original data will lead to the difference in the degree of influence of the array interference phenomenon on different pages. The difference is very significant locally, which seriously affects the accuracy of the prediction.

3.2.2. Input Features

The selection and processing of input features determine whether the prediction algorithm can achieve good results in practical applications.

1. Number of P-E Cycles

From the perspective of application scenarios, the number of remaining P-E cycles is a direct indicator of the endurance of the flash memory. Therefore, the number of P-E cycles is the most direct feature of endurance level prediction.

The total endurance of the actual flash memory block is affected by the programming pressure. Factors such as temperature, programming pattern, and idle period time interval will cause differences in programming pressure, resulting in differences in the total endurance of the flash memory block. Process variation can also cause a huge difference in endurance between flash memory particles. Even with the same batch of flash memory particles of the same process, it is impossible to guarantee that all the particles have the same constitution, which results in a large degree of dispersion of the total P-E cycle number between flash memory particles. Thus, the P-E cycle number cannot be used as a single feature for endurance prediction.

2. Raw Bit Errors

RBE measures the degree of unit wear from the perspective of bad block judgment standards. With the increase in the number of P-E cycles, the RBE numbers of each page of the flash memory particle has increased to varying degrees. The dominant reason for the change of RBE numbers is that the interface traps caused by cell wear cause charge escape/combination and cause the threshold voltage to shift.

3. Erasing Duration

Both programming and erasing operations involve the charge tunneling effect. The interface traps caused by the effect will change the electrical parameters of the memory cell, affecting the tunneling efficiency, which indirectly leads to changes in the operating time. The flash programming strategy causes the programming duration to change with the decrease in endurance, but the programming duration as an input feature is not ideal in actual application scenarios due to great differences in different types of pages in the multi-bit memory cell structure.

The erase operation applies a positive pulse on the substrate to initialize all data in the block to an erased state, and also uses threshold voltage verification to determine whether to apply an additional pulse. However, contrary to the programming operation, the interface traps hinder the tunneling of the charge from the storage layer to the substrate and cause the number of erase pulses to increase, which in turn increases the erase duration.

3.2.3. Label

The endurance judgment is related to the ECC error correction algorithm. When the RBE numbers exceeds the upper limit of the algorithm error correction, the endurance will return to zero. In addition, the garbage collection and out-of-place update mechanisms in the SSD controller [22] lead to amplification effects. Therefore, the number of P-E cycles available at the SSD level is much less than the number of P-E cycles available at the flash block level.

According to the endurance criterion described above, the number of P-E cycles is the metric, and the RBE numbers is the criterion. The endurance level prediction model provides a basis for the endurance level evaluation of the SSD wear leveling algorithm. In addition, the endurance level prediction model can be used to warn and mark the flash memory blocks that will become bad blocks. The number of remaining P-E cycles and RBE numbers are both competent for endurance level prediction.

3.3. Optimization Strategy

3.3.1. RBE Preprocessing

Certain inducing factors will increase the RBE numbers of the partial page in the block. In addition, the arithmetic average can weaken the endurance difference caused by the increase of the RBE numbers of the partial page, reflecting the overall level of endurance of the flash memory block. The erase operation acts on all pages in the flash memory block. If the RBE numbers of a page exceeds the upper limit of ECC error correction, the block will be marked as a bad block. Therefore, the maximum value of page RBE is an effective endurance level predictive input feature.

The standard deviation of the RBE numbers between pages can effectively reflect the difference in endurance between pages. The difference in endurance between pages increases as the endurance decreases, which means that the standard deviation of the RBE numbers can reflect changes in endurance. At the same time, when some non-local disturbances cause overall changes in the RBE numbers of pages within a block, the arithmetic mean will be greatly affected. However, the standard deviation describing the degree of difference can well shield these integrity negative effects of disturbances.

Figure 3 shows a statistical graph of the maximum value of the RBE numbers and standard deviation of a certain block of pages as the number of P-E cycles increases. After ignoring the jitter, the figure shows a monotonous upward trend with the number of P-E cycles, which provides significant rules for machine learning algorithms to learn.

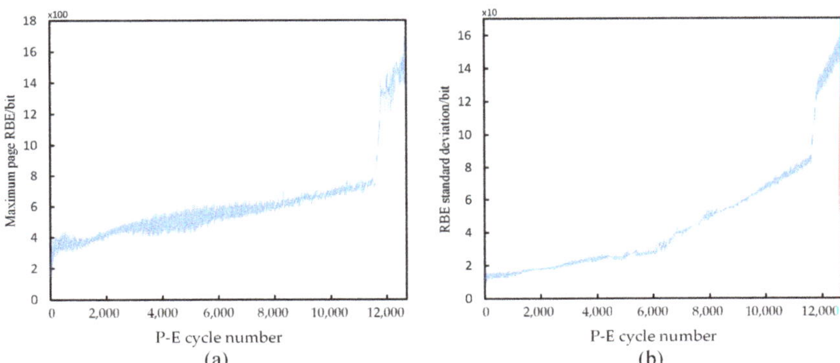

Figure 3. (**a**) The relationship between the maximum RBE numbers of flash memory page and P-E cycle number; (**b**) The relationship between page standard deviation and P-E cycle number.

3.3.2. Transient Error

The transient error caused by the RTN phenomenon has uncertainty: the uncertain drain current causes the threshold voltage to shift randomly, resulting in an uncertain change in the page RBE numbers. Therefore, the page RBE numbers obtained by the test jitters violently as the number of P-E cycles N_c increases, causing significant noise. Since the page RBE numbers is small at the initial stage of wear, and the amount of page RBE change caused by the increase in N_c is not significant, this kind of jitter noise has a great negative impact on endurance level prediction accuracy.

Repeating the P-R-E cycle for a predefined duration and taking the average value of the page RBE can effectively reduce the page RBE numbers noise caused by the RTN phenomenon. However, continuous read operation cannot be used instead of continuous P-R-E operation. Performing a continuous read operation after a single programming operation can ensure that the page RBE numbers obtained each time is under the same N_c. However, in multiple consecutive reading operations, the reading disturbance makes each reading operation affect the result of subsequent reading, and the multiple reading operations are not independent. After the erase operation, the memory cell is approximately restored to the same state, so each P-R-E cycle can be considered independent and does not affect each other. In addition, the RTN phenomenon mainly occurs during the programming operation and for a period of time after it. The RTN phenomenon does not significantly disturb the threshold voltage during multiple continuous read operations. Therefore, the continuous read operation after programming does not improve the negative impact of the RTN phenomenon. Because the maximum value operation is a nonlinear transformation, the sequence of page RBE numbers preprocessing and transient error optimization strategy will have a potential impact on prediction accuracy.

4. Experiments and Analysis

4.1. Test Platform

4.1.1. Test Platform Architecture

A scheme is realized by the Xilinx ZYNQ-7000 series xc7z030ffg676-2 SoC chip (hereafter referred to as ZYNQ-7030) to build a NAND Flash test platform. The test platform consists of a host computer and multiple test boards, as shown in Figure 4. A graphical user interface (GUI) test program runs in the host computer, and multiple test boards are controlled by USB transmission. There are eight test sockets on the test board, and eight BGA132/152 packaged flash memory particles can be tested in parallel at the same time.

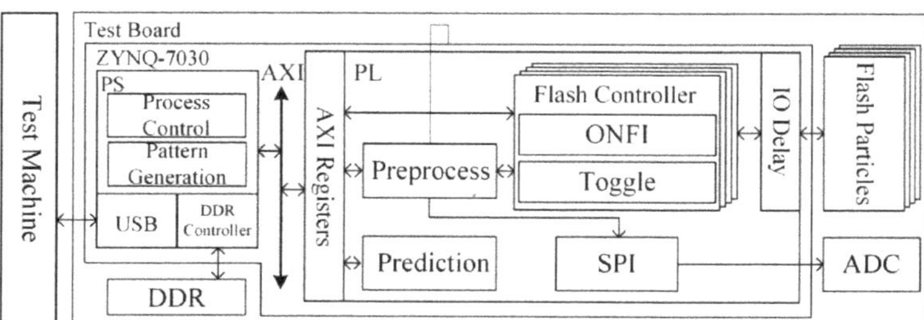

Figure 4. Overview of the Flash test platform.

The test machine is responsible for flash memory specification setting, test process setting, and data storage. Each test board has a ZYNQ-7030 chip, which exchanges data with eight test flash memory particles through GPIO. ZYNQ-7030 can be divided into Processing System (PS) running firmware and Programming Logic (PL) based on Kintex-7 FPGA. The core of the PS is dual-core Cortex-A9, which is mainly responsible for flash memory particle initialization, test flow control, programming pattern drawing, etc. The firmware will automatically generate a test process loop according to the test process parameters transmitted by the host computer, and control the PL-side Flash interface protocol controller module to complete the corresponding command operations through the AXI bus register.

The PL is mainly responsible for functions such as flash interface control, test data processing, and endurance class prediction, including Flash interface protocol controller module, input acquisition and preprocessing module, machine learning prediction algorithm module, SPI protocol control module, etc. There are eight independent Flash interface protocol controller modules, each of which controls the flash memory particles of the channel. The controller module interacts with the PS firmware through the AXI bus register, and the information obtained during the test will pass through the input acquisition and preprocessing module, and then be transmitted to the PS. The SPI protocol control module controls the external 16 bit resolution ADC chip, which can obtain the current value of the flash memory particles at any time and calculate the instantaneous power consumption based on this.

4.1.2. Test Platform Cost

The test platform can test 64×8 flash memory particles at the same time. For a single flash memory particle with a block size of 1024, 50,000 P-E cycles only need 1213 min, and 5 repeated PRE cycles for 1000 blocks only need 261 min. The former corresponds to the data acquisition of the model building process, and the latter corresponds to the data acquisition of the actual application process.

In terms of predictive circuit modules, the use of PL-side FPGA can greatly shorten the time-consuming prediction of endurance levels. After testing, a single prediction of the prediction module under a 100 Mhz clock requires only about 37 us, while a single prediction implemented by PS-side embedded programming requires 108 us.

In terms of resource consumption, the PL-side FPGA hardware resource occupancy is shown in Table 1. The test platform and its proportion are calculated in the case of a 32-bit wide arithmetic unit. In order to achieve highly parallel testing, a single ZYNQ-7030 chip has eight channels internally instantiated, and each channel is divided into four parallel modules to realize multi-CE embedded testing. Each way, the parallel module supports three kinds of interface protocols. In the prediction circuit module, the CORDIC calculation module occupies a higher number of LUTs and Registers resources, which occupy 3925 and 3904, respectively. When the bit width of the calculation unit is reduced to 16 bit, the related resource consumption of the CORDIC calculation module drops significantly to 1159 and 1148.

Table 1. PL Hardware Resource Cost.

Resource	Test Plat-Form (32 bit)	Prediction Circuit		Ratio (32 bit)
		32 bit	16 bit	
Slice LUTs	34,199	5947	1701	44%
Slice Registers	33,732	7689	1941	21%
RAMB36E1	142	28	14	54%
RAMB18E1	9	0	0	2%
DSP48E1s	20	20	8	5%

4.2. Experiment Methodology

4.2.1. Experiment Method

The flash memory particles selected in the experiment are the same batch of MT29F256G08EBHAFJ4 (NW911) chips from Micron. The flash memory particle type of this model is 3D TLC, whose block size is 2304, and the page capacity is 18,588 bytes. In the experiment, the number of consecutive P-R-E cycles $T_\varepsilon = 5$, $T_\alpha + T_\varepsilon = 50$, a set of sample data can be obtained every 50 times of programming, and a total of 96 flash memory blocks are tested for each endurance stage. The programming pattern adopts a PS/PL mixed pseudo-random pattern.

The experiment uses a multi-classification model. There are four labels in the output result of this experiment, which are the maximum level of RBE on the page after 100, 200, and 500 P-E cycles and the level of the number of remaining P-E cycles, which are marked as labels 1–4 in order. As shown in Table 2, each label contains four categories:

- (a) When the number of P-E cycles is used as the dimension, the level of the remaining P-E cycles is used as the output. The category boundaries between different levels are 500, 2500, and 4500 remaining cycles. The four levels of the remaining P-E cycles are $[0, 500), [500, 2500), [2500, 4500), [4500, \infty)$, recorded as level 1–4.
- (b) When RBE is the dimension, the level of the maximum page RBE when the number of P-E cycles $N_c = N_c' + N_i$ is output as the sample point when $N_c = N_c'$, and the category boundaries are 400, 700, and 1000 bits. The four levels of the maximum value of the page RBE are $[0, 400), [400, 700), [700, 1000), [1000, \infty)$, marked as level 1–4. The values of N_i are 100, 200, 500, and there are three tags depending on the difference of N_i.

Table 2. The labels and categories.

Labels	Category 1	Category 2	Category 3	Category 4
1 (P-E cycles)	$[0, 500)$	$[500, 2500)$	$[2500, 4500)$	$[4500, \infty)$
2 (RBE when $N_i = 100$)	$[0, 400)$	$[400, 700)$	$[700, 1000)$	$[1000, \infty)$
3 (RBE when $N_i = 200$)	$[0, 400)$	$[400, 700)$	$[700, 1000)$	$[1000, \infty)$
4 (RBE when $N_i = 500$)	$[0, 400)$	$[400, 700)$	$[700, 1000)$	$[1000, \infty)$

The experiment uses SVM algorithm for model training by default, and the verification method uses five-fold cross-validation. We conducted four independent model trainings. Each training is based on the training data set of one label, and the training data of the remaining three labels is discarded. Since the total number of P-E cycles of the sample flash memory block is wide, and the maximum RBE value at the initial stage of life is generally higher than 300, it is very close to the category boundary 400 of category 1 and category 2. This phenomenon leads to an imbalance in the number of samples in different categories for each label, and the imbalance in the number of samples in different categories is not consistent on different labels. Therefore, before four independent model trainings, we balance the number of samples of different categories for each label, and reduce the number of samples of the three larger categories to the smallest number of category samples by random selection—the total number of samples for label 1 was found to be 6576, label 2 was 8272, label 3 was 8952, and label 4 was 15,208.

4.2.2. Evaluation Indicators

1. Confusion Matrix

The confusion matrix is a visual numerical matrix used to reflect the classification results of a supervised machine learning model. Various indicators of the classifier model are calculated based on the confusion matrix. The confusion matrix of the L classifier model is a square matrix of $L \times L$, which can intuitively reflect the distribution of each actual category and output category. Each row of the confusion matrix belongs to the same actual category, and each column belongs to the same output category.

2. Numerical Indicators

The two-class model commonly uses Accuracy (A), Precision (P), Recall (R), and F1-score (F_1) to measure the pros and cons of the model. However, in the multi-class model, the increase in the number of rows and columns of the confusion matrix leads to ambiguous indicator definitions, so corresponding changes are needed. The expression and meaning of the numerical indicators of the classification model are shown in Table 3. Among them, the accuracy and recall rate in the multi-classification model are divided into three categories: macro, micro, and weighted, and the Kappa coefficient is introduced.

Table 3. The expression and meaning of the numerical indicators of the classification model.

Indicators	Two-Class	Multi-Class	Meaning
Accuracy	$\frac{TP+TN}{TP+TN+FP+FN}$	$\frac{1}{S} \cdot \sum_m A_{m,m}$	Overall prediction accuracy rate
Precision	$\frac{TP}{TP+FP}$	$macro-P = \frac{1}{L}\sum_n \frac{A_{n,n}}{\sum_m A_{m,n}}$ $micro-P = Accuracy$ $weighted-P = \sum_n \frac{A_{n,n} \cdot \sum_m A_{n,m}}{S \cdot \sum_m A_{m,n}}$	Ratio of correct predicted value
Recall	$\frac{TP}{TP+FN}$	$macro-R = \frac{1}{L}\sum_m \frac{A_{m,m}}{\sum_n A_{m,n}}$ $micro-R = weighted-R = A$	Ratio of correct true value
F1-score	$\frac{2P \cdot R}{P+R}$	$\frac{2P \cdot R}{P+R}$	Precision/recall rate trade-off value
Kappa		$\frac{S \cdot \sum_m A_{m,m} - \sum_n (\sum_m A_{m,n})(\sum_m A_{n,m})}{S^2 - \sum_n (\sum_m A_{m,n})(\sum_m A_{n,m})}$	Biased consistency indicators

3. ROC Curve

The ROC curve is often used for model comparison and threshold screening in the case of classification. The AUC value of the area under the curve can intuitively reflect the pros and cons of performance. The larger the AUC value, the better the performance. In a multi-class model, each category corresponds to a ROC curve, and it is necessary to ensure that the categories are the same when comparing models.

In summary, when comparing the prediction results, we will compare the accuracy rate A, macro accuracy $macro-P$, macro recall rate $macro-R$, *macro*-$F1$ score, Kappa coefficient K, and roc curve.

4.3. *Analysis*

4.3.1. Comparison of the Results of Different Labels

This experiment uses Binary Relevance technology to transform the multi-label multi-classification model into L single-label multi-classification models. Each label is trained separately to obtain the prediction result, and the results of different labels are compared. The numerical indicator results are shown in Table 4.

Table 4. Statistical table of model numerical indicators of different labels.

Labels	A	$macro-P$	$macro-R$	$macro-F_1$	K
1	0.958222	0.958532	0.958222	0.958377	0.944296
2	0.952369	0.953216	0.952369	0.952793	0.936493
3	0.939173	0.940106	0.939173	0.939639	0.918897
4	0.897679	0.897441	0.897679	0.897561	0.863572

The numerical indicators of label 1 are the best. The first four indicators all reach 95.8%, the Kappa coefficient is about 94.4%, and the K value greater than 90% means that the model has extremely high consistency. The effect of the numerical indicators of label 2 is followed closely. The first four indicators are about 95.2–95.3%; the gap is not big. The prediction effect of the model of label 4 is the worst. The first four indicators are about 89.7%, and the K value is only 86.4%.

Figure 5a visually compares the correct rate A and Kappa coefficient K of the models of different labels through the bar graph. The difference between labels 1 to 3 is not big, label 1 is the best, and label 4 is obviously different. The indicator gap between the four labels is related to the classification basis. Labels 1 to 3 are divided into categories based on the RBE numbers, which essentially predicts the changes of certain parameters after a certain number of times in the future. At the same time, the difference between tags 1 to 3 is that the value of N_i is different. In addition, the value of N_i reflects the number of times the predicted target is away from the current state. Tag 1 has the smallest gap and tag 3 has the largest. The smaller the gap means the smaller the change based on the current state, and the higher the accuracy of the prediction will naturally be. Tag 4 is divided into categories based on the number of P-E cycles. Essentially, it judges the current endurance stage based on the characterization of the current endurance parameters, and it also needs to determine the total endurance range. The large difference in endurance and mischaracterization between the flash memory particles greatly weakens the model's ability to judge.

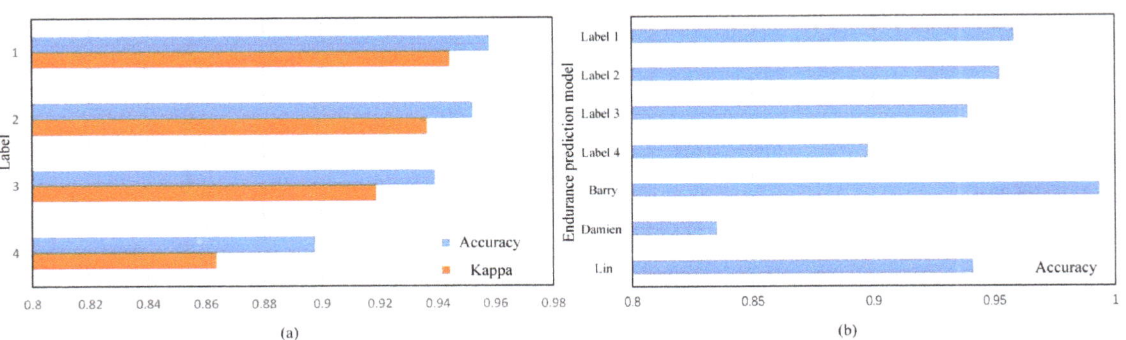

Figure 5. (**a**) Comparison of model accuracy and Kappa coefficient of different labels; (**b**) Comparison of the accuracy of models in other studies.

The comparison between the models corresponding to the four labels and the endurance prediction models of other researchers is shown in Figure 5b. Barry's scheme worked best, achieving a 99.4% correct rate. However, the number of negative samples in the study only accounted for 0.03%, which greatly reduced its reliability. The correct rates of labels 1–3 and Lin's models in this scheme are about 94–96%, followed by label 4. These models are ahead of the 83.5% correct rate of Damien's scheme. Excluding the unreliable Barry scheme due to extremely unbalanced samples, the model accuracy rate of this scheme is in the first echelon in this field. Compared with the two-class judgment of other schemes, this scheme adopts a multi-classification model, and the increase in the number of categories makes it more abundant in application scenarios. In addition to basic bad block warning, the model of this solution can also be used for wear leveling strategies, factory screening and rating, etc.

Since the AUC value of category 4 of each model is about 0.99, the upper left corner area is enlarged to the lower right corner for display. As shown in Figure 6a, the AUC value relationship of tags 1–3 is consistent with the correct rate relationship, that is, $AUC_1 > AUC_2 > AUC_3$.

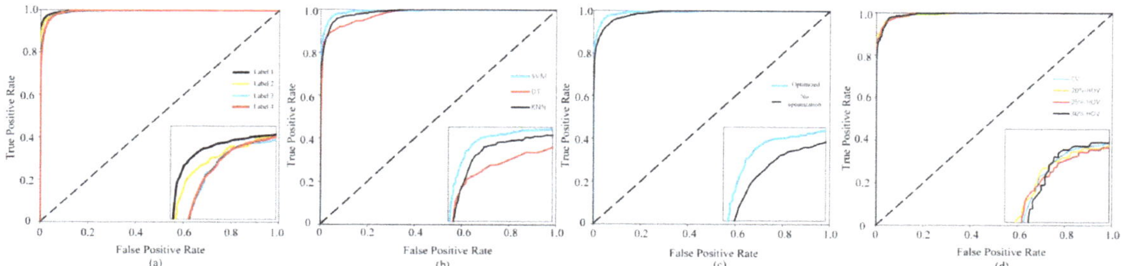

Figure 6. (**a**) ROC curves of models with different labels; (**b**) ROC curves of different algorithms; (**c**) ROC curves of models with or without transient error optimization; (**d**) ROC curves of models with different verification method.

The ROC curve of category 4 of labels 1 to 3 reflects the model's prediction of bad blocks, because the boundary of category 4 is close to the critical value of bad block judgment. This shows that the pros and cons of the bad blocks predicted by tags 1–3 are also consistent with the overall pros and cons of the model. However, the special case is that the ROC curves of label 4 and label 3 are very close, but there is a gap between the two in numerical indicators. There are two main reasons for this phenomenon: first, the classification dimensions of label 3 and label 4 are not consistent, and the meaning of category 4 is not the same. It is meaningless to directly compare the ROC curves of the two categories 4; second, the selected numerical indicators. It reflects the overall situation of the four categories, and there are differences between the local and the whole. In fact, the correct rate A of each category is calculated separately, and the correct rates A of label 3 and label 4 to category 4 are 96.27% and 96.26%, respectively, which are very close. However, the correct rate A of label 3 for categories 2 and 3 is 97.64% and 95.26%, while the correct rate A of label 4 for categories 2 and 3 is only 93.55% and 92.18%, which is a large gap.

Considering the evaluation indicators and actual application scenarios, this paper believes that the prediction model of label 3 is the best because the numerical indicator of label 4 is low. When the difference between the internal numerical indicators of labels 1 to 3 is not large, the value of N_i of label 3 is larger, which means that label 3 can make early warning and decision-making in actual application scenarios. Therefore, when comparing other variables in the follow-up, they will all be discussed in the case of label 3.

4.3.2. Comparison of Results of Different Algorithms

In addition to the default SVM algorithm, we also use the DT algorithm and the KNN algorithm to perform model training on the same training set. The results are shown in the Table 5. Figure 7 shows Accuracy and Kappa coefficient of different algorithms. The SVM algorithm has achieved the best results in all the numerical indicators in the table, which is about 3% to 4% higher than the DT algorithm, and about 5% to 8% higher than the KNN algorithm. At the same time, this experiment has conducted sample balance processing between each category, considering the KNN algorithm's classification disadvantages of unbalanced sample data sets; in actual situations, when the endurance level prediction scheme is applied, the KNN algorithm may be more disadvantageous in accuracy.

Table 5. Statistical table of model numerical indicators of different algorithms.

Labels	A	$macro-P$	$macro-R$	$macro-F_1$	K
SVM	0.939173	0.940106	0.939173	0.939639	0.918897
DT	0.907391	0.909162	0.907391	0.908275	0.876521
KNN	0.880779	0.889564	0.880779	0.885149	0.841038

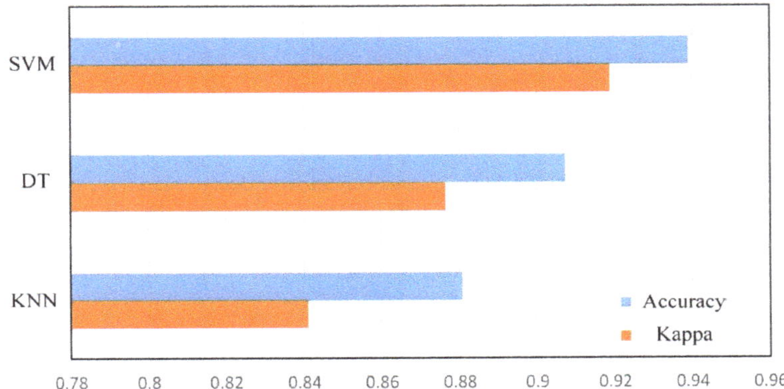

Figure 7. Comparison of model accuracy and Kappa coefficient of different algorithms.

It can be seen from the ROC curve in Figure 6b that the SVM algorithm is still the best in performance of the AUC value, while the DT algorithm is the worst, and the gap is obvious. Because the ROC curve for category 4 reflects the model's classification performance in the critical value of bad block judgment, and is an important indicator of the pros and cons of the bad block early warning function, the DT algorithm has a great disadvantage in this important function.

4.3.3. Analysis of Transient Error Optimization Effect

1. Comparison of Optimized and No Optimization

Comparing the effect of optimization with or without transient errors will inevitably lead to an imbalance in the number of category samples in one of the cases. Therefore, the weighted-P indicator is added to the statistical table of numerical indicators of the model with or without optimization, as shown in Table 6.

Table 6. Statistical table of numerical indicators of optimization and no optimization.

Labels	A	$macro-P$	$macro-R$	$macro-F_1$	K
Optimized	0.939173	0.940106	0.939173	0.939639	0.940106
No optimization	0.860097	0.8575992	0.858039	0.857819	0.864401

The input of the non-optimized model takes the last time of the T_α cycle, and the input of the optimized model uses the first average processing method. From a comparison of Table 6, it can be seen that there is a huge gap in the numerical indicators of the model with or without transient error optimization. The accuracy rate A of the optimized model is 7.9% higher than that of the non-optimized model. Among the other five indicators, the optimized model is about 7% to 10% higher than the non-optimized model, which is a significant improvement. This is because the transient error optimization strategy can significantly reduce the jitter noise in the endurance data, so that the machine learning algorithm can better analyze the intrinsic relationship between the endurance level and the input vector.

The ROC curve in Figure 6c shows that the AUC value of the optimized model is higher than that of the non-optimized model, so the optimized model judges bad blocks more accurately. The comparison result fully illustrates the necessity and correctness of the transient error optimization strategy.

2. Comparison of Optimization Order

The sequence problem of the maximum/standard deviation operation and the transient error optimization operation will cause the difference of the input vector after the transient error optimization, which is essentially caused by the characteristics of the nonlinear transformation. The previous forecasting models are all optimized before processing. Table 7 shows the numerical indicator results of earlier transient optimization and later transient optimization.

Table 7. Statistical table of numerical indicators of optimization order.

Labels	A	$macro-P$	$macro-R$	$macro-F_1$	K
Pre-optimization	0.939173	0.940106	0.939173	0.939639	0.940106
Post-optimization	0.915602	0.915280	0.917297	0.916288	0.916810

It can be seen from the table that pre-optimization is better in the prediction results. The accuracy rate A has achieved a lead of 2.4%, and the other indicators have achieved a lead of 2% to 3%. The result is related to the theoretical basis of the optimization strategy. The transient error optimization strategy is based on the theoretical situation that probability $p = \psi(N_c)$ can be considered as a fixed value, when N_c is approximately constant. However, the function $\varphi(N_c)$ of different storage units is different, owing to which the theoretical situation applies only to the same storage unit or page. Post-optimization will cause the optimization strategy to deviate from the theoretical situation. At the same time, in the early stage of endurance when the page RBE numbers change little, the gap between the RBE numbers pages is little. The errors caused by various disturbance factors account for a relatively large amount. The post-optimization will greatly weaken the effect of transient error optimization. The pre-optimization can ensure that the input vector of $f(S_k)$ comes from the same flash page, so that the theoretical situation is applicable and no negative effects will occur. Therefore, the pre-optimization method can achieve better optimization results.

4.3.4. Analysis of Validation Method and Feature Correlation

1. Comparison of Different Validation Methods

The performance of classification model results is evaluated and compared in a five-fold cross-validation method. The advantage of this method is to reduce the statistical uncertainty of the average test error estimation, so as to facilitate model comparison and result analysis. In order to avoid misjudgments caused by differences in validation methods, we also compared the prediction results of the five-fold cross-validation and different ratios of Hold-Out validation methods. The Hold-Out ratios are selected as 20%, 25%, and 30%.

According to the comparison results of the indicators in Table 8, the best indicator is the Hold-Out method with a ratio of 20%, and the worst is the Hold-Out method with a ratio of 30%. The accuracy difference between the two is about 1.7%. The Kappa coefficient gap is about 2.3%. In fact, the prediction result of the Hold-Out method changes greatly due to the difference in the selection of the test set. Taking the 20% ratio Hold-Out method as an example, the accuracy rates A of the five repetitive training models with the same data set are 94.52%, 92.99%, 93.83%, 93.60%, and 94.66%, and the difference between the maximum and minimum values is about 1.67%. Taking into account the indicator fluctuations caused by the difference in the selection of the test set, it can be considered that the numerical indicators of the models in the four cases are very close.

Table 8. Statistical table of numerical indicators of different validation methods.

Labels	A	$macro-P$	$macro-R$	$macro-F_1$	K
Five-fold cross-validation	0.939173	0.940106	0.939173	0.939639	0.918897
20% Hold-Out	0.945247	0.945659601	0.94525632	0.945457917	0.92699593
25% Hold-Out	0.934307	0.936912156	0.934306569	0.935607549	0.912408759
30% Hold-Out	0.927992	0.929085804	0.927991886	0.928538523	0.903989182

The ROC curve in Figure 6d also confirms this conclusion. The ROC curves of the four models are very close. Therefore, the Hold-Out method with a separate test set can still obtain almost the same evaluation index, indicating that the model obtained by the endurance level prediction scheme can still achieve excellent prediction results in the additional test set.

2. Feature Correlation

At present, the features of the experiment are the arithmetic mean, maximum, and standard deviation of the page RBE numbers, as well as the number of P-E cycles and the duration of erasure. When performing feature analysis, the Pearson correlation coefficient r can be used to measure the linear correlation between the various dimensions of the input vector. When its value is close to 1, it means that the redundancy of the input vector space is large, and the dimension of the input vector can be simplified. Through calculation, the Pearson correlation coefficient $r(RBE_a, RBE_s)$ between the arithmetic mean of the page RBE numbers and the standard deviation is:

$$r(RBE_a, RBE_s) = \frac{\sum_i (X_i - \overline{X})(Y_i - \overline{Y})}{\sqrt{\sum_i (X_i - \overline{X})^2}\sqrt{\sum_i (Y_i - \overline{Y})^2}} = 0.9790 \quad (1)$$

The value is extremely close to 1, which means that there is a strong linear correlation between the two vectors. When the two are used as the model input vector dimensions at the same time, the recognition of the feature relationship between the input and the output is of minimal help. At the same time, the prediction circuit needs to perform parallel calculations on various dimensions of input. If the input dimensions can be reduced, the hardware resource consumption will be greatly reduced.

As shown in Table 9, the PCA dimensionality reduction method reduces the input from five dimensions to four. Through comparison, it can be found that the complete input still achieves the best prediction effect, but the lead is extremely small. Excluding the two cases of RBE arithmetic mean/standard deviation, the difference is too small to be ignored. Considering that the arithmetic mean can shield the local disturbance, and the standard deviation can shield the overall disturbance, the linear correlation between the two vectors is extremely strong, indicating that the impact of the endurance change on the overall disturbance and the local disturbance is positively correlated.

Table 9. Statistical table of numerical indicators of input dimension reduction.

Labels	A	$macro-P$	$macro-R$	$macro-F_1$	K
Five inputs	0.939173	0.940106	0.939173	0.939639	0.918897
No arithmetic mean	0.934002	0.935897511	0.934002433	0.934949012	0.912003244
No standard deviation	0.933394	0.935000947	0.933394161	0.934196863	0.911192214
PCA method	0.937196	0.939040182	0.937195864	0.938117116	0.916261152

Figure 8 shows that the arithmetic mean and standard deviation are roughly linear distributions, which are consistent with the results. The model indicator using the PCA dimensionality reduction method is very close to the complete input model indicator, but this method requires dimensionality reduction through a certain function transformation, which will add additional hardware resources. The method of removing the arithmetic mean/standard deviation of the page RBE reduces the consumption of hardware resources on the premise that the difference between the arithmetic mean/standard deviation and the complete input is negligible. Therefore, the input of the standard deviation will be removed in the implementation of the specific scheme.

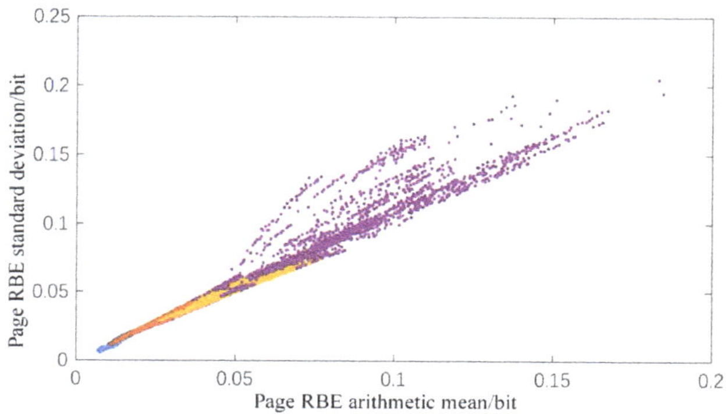

Figure 8. The distribution of the arithmetic mean and standard deviation of the sample points of the data set.

4.4. Application

The endurance level prediction model has many practical application scenarios. The paper designs a simple warning strategy for bad blocks based on the prediction scheme introduced. In the actual application scenario of bad block warning, the prediction model will face the problem of recall rate. Assuming that a block will become a bad block after a certain number of programming times, the recall rate determines the probability that the prediction model can be used to successfully judge and give an early warning. The recall rate is the most important evaluation indicator in the early warning of bad blocks. In data-sensitive fields, users stop using the flash memory when the usage of flash memory reaches half of the nominal value because the method can make the recall rate reach 100%. Even if the method will cause the real usage rate of flash memory to be much lower than 10%, it is necessary to ensure that no bad blocks are missed. Therefore, the prediction model applied to the bad block early warning strategy needs to achieve the following goals:

(1) Improve the recall rate of the bad block warning strategy as much as possible.
(2) On the premise of ensuring goal 1, try to increase the real utilization rate of flash memory, that is, postpone the bad block warning time.
(3) Reduce the wake-up frequency of the prediction program and prediction circuit to reduce the burden on the SSD controller.

Based on the above objectives, this paper designs a comprehensive strategy for early warning of bad blocks. Take Figure 9 as an example—the curve in the figure represents a schematic curve of the error rate of the flash memory block as a function of the number of P-E cycles. Assuming that the flash memory reaches the critical value of bad blocks at point N, an uncorrectable data error occurs. The number of P-E cycles between point M and point N differs by 500. Point P is the first time that the predictive model judges the block to be a positive type (assuming that Bad block) moment. Before point P, the bad block warning strategy wakes up the endurance level prediction circuit during every A programming operation. After the P point, it is changed to wake up once every B programming operation,

B < A. In this way, the prediction circuit can be called with a lower wake-up frequency during the endurance stage with lower risk, and frequent predictions when approaching the end of the endurance period, in order to take into account resource consumption and the accuracy of the early warning strategy. In the judgment of bad blocks, the category 4 of the prediction model labels 1 to 3 or category 1 of the label 4 are regarded as the positive type, because they both mean that the prediction result of the flash memory block is located at point M and to the right. The prediction circuit is frequently awakened after point P. If C consecutive prediction results are positive, an early warning is sent to the controller.

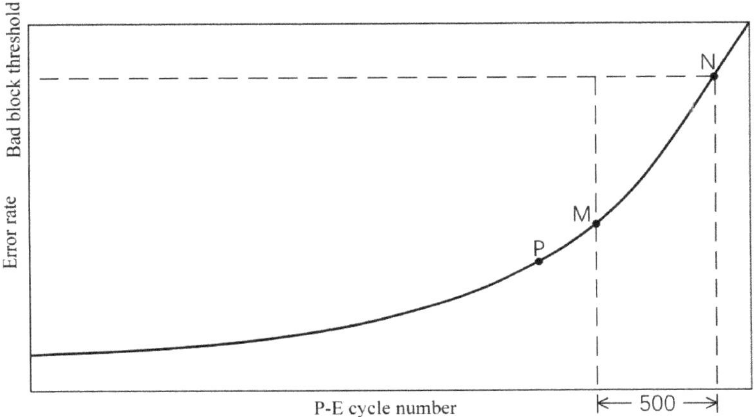

Figure 9. Schematic diagram of the flash block error rate and the number of P-E cycles.

Let A be 200 and B be 50, and use the prediction model of label 3 to test. After testing 96 sample blocks, the program successfully provided early warning for 93 blocks, with a success rate of about 96.9%. The recall rate of the model category 4 is only 89.90%, which shows that the bad block early warning strategy can successfully improve the accuracy of the endurance class prediction model in practical applications under the condition of low wake-up rate.

5. Conclusions

In order to effectively prolong the service life of flash memory and avoid the loss caused by sudden failure, this paper conducts related research on flash memory endurance, proposes a flash memory endurance grade prediction scheme based on the SVM algorithm, and designs a high parallel test platform and low time-consuming endurance prediction module based on FPGA. We research and analyze the feature quantities closely related to the endurance changes in the flash memory, and determine that the model takes the block as the object. The page RBE numbers, the number of P-E cycles, and the erase duration in the block are used as the input feature quantity, and the output is the remaining lifetime level, or RBE numbers level after 100/200/500 P-E cycles. This scheme adopts a variety of strategies to reduce the negative interference in the forecasting process in a targeted manner. The prediction module is realized based on the ZYNQ-7030 chip. The SVM decision model is deconstructed and the parallel multiplication structure is designed to realize the highly multiplexed pipelined calculation. The prediction module only needs 37 us per time, which greatly reduces the time consumption of prediction.

The method uses multi-category evaluation indicators to analyze five aspects: four tags achieved 89.77–95.82% accuracy, each evaluation indicator is in the leading echelon, and the increase in the number of categories expands the scope of application. Compared with DT and KNN, the SVM model of the RBF kernel function achieved a lead of 3–8%. The model using the transient error optimization strategy achieved an indicator increase of 7–

10%, and pre-optimization leads up to 2% to 3%. Cross-validation and Hold-Out validation results show that the model can still achieve the same prediction effect in the additional test set. Pearson correlation coefficient analysis shows that the impact of the endurance change on the overall disturbance and the local disturbance is positively correlated. Finally, the bad block early warning strategy designed based on the proposed model can successfully achieve early warning for 96.9% of the blocks.

Author Contributions: The work presented here was completed in collaboration between all authors. H.Z. prepared the manuscript, designed the test platform, and performed some of the experiments. J.W. developed the test platform and also performed some of the experiments. Z.C. developed the test platform and also performed some of the experiments. Y.P. performed some of the experiments as well. Z.L. (Zhaojun Lu) performed some of the experiments and revised the manuscript. Z.L. (Zhenglin Liu) proposed the ideas and revised the manuscript. All authors have read and agreed to the published version of the manuscript.

Funding: This work was supported in part by the National Natural Science Foundation of China (Grant No. 61874047) and a grant of key technologies R&D general program of Shenzhen, No. 20201102 3000308.

Conflicts of Interest: The authors declare that there is no conflict of interest.

References

1. IC Insights. Memory Market Not Forecast to Exceed 2018 High of $163.3B until 2022. Available online: https://www.icinsights.com/news/bulletins/Memory-Market-Not-Forecast-To-Exceed-2018-High-Of-1633B-Until-2022 (accessed on 25 May 2020).
2. Lee, S.Y.; Schroder, D.K. 3D IC architecture for high density memories. In Proceedings of the 2010 IEEE International Memory Workshop, Seoul, Korea, 16–19 May 2010; pp. 1–6.
3. Alsalibi, A.I.; Mittal, S.; Al-Betar, M.A.; Sumari, P.B. A survey of techniques for architecting SLC/MLC/TLC hybrid Flash memory-based SSDs. *Concurr. Comput. Pract. Exp.* **2018**, *30*, e4420.1–e4420.21. [CrossRef]
4. Xia, Z.; Kim, D.S.; Jeong, N.; Kim, Y.G.; Kim, J.H.; Lee, K.H.; Park, Y.K.; Chung, C.; Lee, H.; Han, J. Comprehensive modeling of NAND flash memory reliability: Endurance and data retention. In Proceedings of the 2012 IEEE International Reliability Physics Symposium (IRPS), Anaheim, CA, USA, 15–19 April 2012; pp. 1–4.
5. Deguchi, T.N.Y.; Takeuchi, K. 9.1x Error acceptable adaptive artificial neural network coupled LDPC ECC for charge-trap and floating-gate 3D-NAND flash memories. In Proceedings of the 2018 IEEE Custom Integrated Circuits Conference (CICC), San Diego, CA, USA, 8–11 April 2018; pp. 1–4.
6. Micron. Wear-Leveling Techniques in NAND Flash Devices. Technical Note TN-29-42. 2008, pp. 1–8. Available online: https://www.icinsights.com/news/bulletins/Memory-Market-Not-Forecast-To-Exceed-2018-High-Of-1633B-Until-2022 (accessed on 27 March 2021).
7. Desnoyers, P. Empirical Evaluation of NAND Flash Memory Performance. *Oper. Syst. Rev.* **2010**, *44*, 50–54. [CrossRef]
8. Pan, Y.; Zhang, H.; Gong, M.; Liu, Z. Process-variation effects on 3D TLC flash reliability: Characterization and mitigation scheme. In Proceedings of the 2020 IEEE 20th International Conference on Software Quality, Reliability and Security (QRS), Macau, China, 11–14 December 2020; pp. 1–6.
9. Hogan, D. Genetic Programming Based Predictions and Estimations for the Endurance and Retention of NAND Flash Memory Devices. Ph.D. Thesis, University of Limerick, Limerick, Ireland, 2013.
10. Fitzgerald, B.; Hogan, D.; Ryan, C.; Sullivan, J. Endurance prediction and error Reduction in NAND flash using machine learning. In Proceedings of the 2017 17th Non-Volatile Memory Technology Symposium (NVMTS), Aachen, Germany, 30 August–1 September 2017; pp. 1–8.
11. Fitzgerald, B.; Fitzgerald, J.; Ryan, C. A comparative study of classification methods for flash memory error rate prediction. In Proceedings of the International Conference on Advanced Machine Learning Technologies and Applications, Cairo, Egypt, 22–24 February 2018; pp. 385–394.
12. Ma, R.; Wu, F.; Zhang, M.; Lu, Z.; Wan, J.; Xie, C. RBER-Aware lifetime prediction scheme for 3D-TLC NAND flash memory. *IEEE Access* **2019**, *7*, 44696–44708. [CrossRef]
13. Micheloni, R.; Crippa, L.; Marelli, A. *Inside Nand Flash Memories*; Springer: Berlin, Germany, 2010; pp. 19–20.
14. Lee, J.D.; Lee, C.K.; Lee, M.W.; Kim, H.S.; Park, K.C.; Lee, W.S. A New Programming Disturbance Phenomenon in NAND Flash Memory By Source/Drain Hot-Electrons Generated By GIDL Current. In Proceedings of the 2006 21st IEEE Non-Volatile Semiconductor Memory Workshop, Monterey, CA, USA, 12–16 February 2006; pp. 31–33.
15. Chimenton, A.; Zambelli, C.; Olivo, P. A Statistical Model of Erratic Behaviors in Flash Memory Arrays. *IEEE Trans. Electron Devices* **2011**, *58*, 3707–3711. [CrossRef]
16. Smith, F.W. Pattern Classifier Design by Linear Programming. *IEEE Trans. Comput.* **1968**, *100*, 367–372. [CrossRef]

17. Vapnik, V.N.; Chervonenkis, A.Y. On the uniform convergence of relative frequencies of events to their probabilities. In *Measures of Complexity*; Springer: Berlin/Heidelberg, Germany, 2015; pp. 11–30.
18. Hsu, C.W.; Lin, C.J. A comparison of methods for multi-class support vector machines. *IEEE Trans. Neural Netw.* **2002**, *13*, 415–425. [PubMed]
19. Cai, Y.; Haratsch, E.F.; Mutlu, O.; Mai, K. Error patterns in MLC NAND flash memory: Measurement, characterization, and analysis. In Proceedings of the 2012 Design, Automation & Test in Europe Conference & Exhibition (DATE), Dresden, Germany, 12–16 March 2012; pp. 521–526.
20. Hsieh, C.C.; Lue, H.T.; Li, Y.C.; Chen, T.W.; Li, H.P.; Lu, C.Y. A Novel Dichotomic Programming Gorithm Applied to 3D NAND Flash. In Proceedings of the 2015 Symposium on VLSI Technology (VLSI Technology), Kyoto, Japan, 16–18 June 2015; pp. T180–T181.
21. Olson, A.R.; Langlois, D.J. Solid State Drives Data Reliability and Lifetime. *Imation White Pap.* **2008**, 1–27.
22. Hu, X.Y.; Eleftheriou, E.; Haas, R.; Iliadis, I.; Pletka, R. Write Amplification Analysis in Flash-Based Solid State Drives. In Proceedings of the 2009 Israeli Experimental Systems Conference, Haifa, Israel, 4–6 May 2009; p. 10.

Article

A Scalable Bidimensional Randomization Scheme for TLC 3D NAND Flash Memories

Michele Favalli [1,†], Cristian Zambelli [1,*,†], Alessia Marelli [2,†], Rino Micheloni [3,†] and Piero Olivo [1,†]

1 Dipartimento di Ingegneria, Università degli Studi di Ferrara, Via G. Saragat 1, 44122 Ferrara, Italy; michele.favalli@unife.it (M.F.); piero.olivo@unife.it (P.O.)
2 Freelance Consultant, Via Don Pino Puglisi 4, 24048 Treviolo, Italy; alessiamarelli@gmail.com
3 Freelance Consultant, Via Roma 23, 22010 Moltrasio, Italy; rino.micheloni@ieee.org
* Correspondence: cristian.zambelli@unife.it; Tel.: +39-0532-974-993
† These authors contributed equally to this work.

Abstract: Data randomization has been a widely adopted Flash Signal Processing technique for reducing or suppressing errors since the inception of mass storage platforms based on planar NAND Flash technology. However, the paradigm change represented by the 3D memory integration concept has complicated the randomization task due to the increased dimensions of the memory array, especially along the bitlines. In this work, we propose an easy to implement, cost effective, and fully scalable with memory dimensions, randomization scheme that guarantees optimal randomization along the wordline and the bitline dimensions. At the same time, we guarantee an upper bound on the maximum length of consecutive ones and zeros along the bitline to improve the memory reliability. Our method has been validated on commercial off-the-shelf TLC 3D NAND Flash memory with respect to the Raw Bit Error Rate metric extracted in different memory working conditions.

Keywords: 3D NAND Flash; RBER; reliability; flash signal processing; randomization scheme

Citation: Favalli, M.; Zambelli, C.; Marelli, A.; Micheloni, R.; Olivo, P. A Scalable Bidimensional Randomization Scheme for TLC 3D NAND Flash Memories. *Micromachines* **2021**, *12*, 759. https://doi.org/10.3390/mi12070759

Academic Editor: Jung-Ho Yoon

Received: 31 May 2021
Accepted: 25 June 2021
Published: 27 June 2021

1. Introduction

The 3D NAND Flash technology is the primary choice for non-volatile mass storage platforms such as Multimedia Cards (MMCs) and Solid State Drives (SSDs) [1]. Compared with its planar predecessor, this technology offers a significantly higher storage density and better scaling features [2,3]. From the reliability standpoint, the 3D NAND Flash technology inherits the issues already documented for planar NAND Flash, such as wear-out failures due to repeated data writing/erasing (i.e., endurance failures [4]), high temperature sensitivity either in static (i.e., data retention [5]) or in dynamic (i.e., cross-temperature [6]) working conditions, and disturbances due to frequent access to the memory (e.g., read disturb [7]). On top of these, novel reliability threats specifically belonging to the physical nature of 3D devices come into play [8,9]. At the system level, all these reliability detractors are perceived through an increase of the Fail Bits Count (FBC) exposed by the 3D NAND Flash after operation. An efficient way to handle the ever-growing FBC during the entire memory lifetime is to rely on complex Error Correction Code (ECC) engines [10] that work on a translation of the FBC concept into an equivalent Raw Bit Error Rate (RBER) to perform the error recovery. However, the RBER is highly dependent on the pattern applied to write the data on the memory; therefore, without decoupling it from the intrinsic 3D NAND Flash reliability, we would experience some unfortunate situations where RBER is higher than the Shannon's limit [11].

A fundamental component in storage systems, the goal of which is to avoid these events, is the randomizer. This object ensures that the memory data programming is achieved in the most efficient way, making the probability of a worst-case data pattern statistically negligible. The idea behind data randomization is to perform a transformation from original user data by simply inserting an exclusive OR (XOR) operation between the data path and the output of a Linear Feedback Shift Register (LFSR) initialized by a seed [12,13]. The seed is the starting value to be loaded into the LFSR to enable the generation of random patterns. The value of the seed is usually selected to avoid pattern repetitions and must be greater than zero to start the generation of pseudo-random sequences. Additionally, multiple seeds can be exploited to reduce the correlation effects between different LFSRs' random number generations. The randomizer block can be implemented either on-chip in the circuit periphery close to the memory array [14,15], or off-chip by implementing its function in the storage controller when its architectural complexity requires additional logical operations, such as for example, those required for seed generation [16]. Despite the importance of this component, we must note that the focus of these implementations is on the random value patterns' generation primarily in the horizontal dimension (i.e., page- or wordline-wise), while being less effective in the vertical dimension (i.e., string- or bitline-wise) of the memory. Most randomization schemes do not bother with the maximum number of consecutive ones or zeros along the bitline that could impair the sensing operation, thus resulting in a localized RBER increase.

In the literature, cumbersome methodologies based on multiple chained LFSRs, or even on look-up tables exploited for seed generation with arithmetic functions based on heuristics, are adopted [16]. However, all the proposed solutions lack information in terms of the mathematical approach required to achieve randomization. From the storage system designer point of view, this will be a limiting factor since every time there is a technology update of the storage medium (e.g., a change to the memory density, storage paradigm, etc.), there will be a forced change of the randomization scheme.

In this work, we tackle the data randomization challenge in Triple Level Cell (TLC) 3D NAND Flash memories by presenting a simple yet scalable bi-dimensional randomizer that guarantees both the horizontal and the vertical randomization while defining an upper bound on the maximum sequence length of consecutive ones and zeros along the dimensions.

The contributions of this paper can be summarized as follows:

1. We show that a chained structure of two k-bits LFSRs can provide, from a statistical standpoint, both the horizontal and vertical data randomization while guaranteeing a k-bit upper bound on the maximum sequence length of consecutive ones and zeros;
2. We show that our proposed randomization scheme introduces a low-complexity hardware overhead, most of which scales automatically with the memory size and is independent of cumbersome heuristics, to achieve seed randomization or look-up tables (LUTs) for seed storage, to potentially be adopted by different memory technologies and vendors;
3. We benchmark the effectiveness of our scheme by measuring the RBER characteristics of a Triple Level Cell (TLC) 3D NAND Flash memory during both endurance and data retention stress.

2. Background

2.1. 3D NAND Flash Memory Architecture and Randomization Fundamentals

The architecture of a 3D NAND Flash is described in the sketch in Figure 1a. The primary element of the array is the stack of Control Gates (CGs), also indicated as Layers. Associated with each CG, there are several wordlines that depend on the specific integration concept for the memory [17]. The bottom of the memory architecture is represented by the Source Line and the Source Line Selectors of the 3D NAND Flash string. Multiple holes are drilled through the CG stacks and plugged with poly-silicon in order to form a series of vertically arranged 3D NAND Flash memory cells. In TLC architectures, all the cells belonging to a wordline can store up to three bits per cell, defined as Lower Significant Bit (LSB), Central Significant Bit (CSB) and Most Significant Bit (MSB); Bitline selectors and bitline (BL) contacts are on top of the structure.

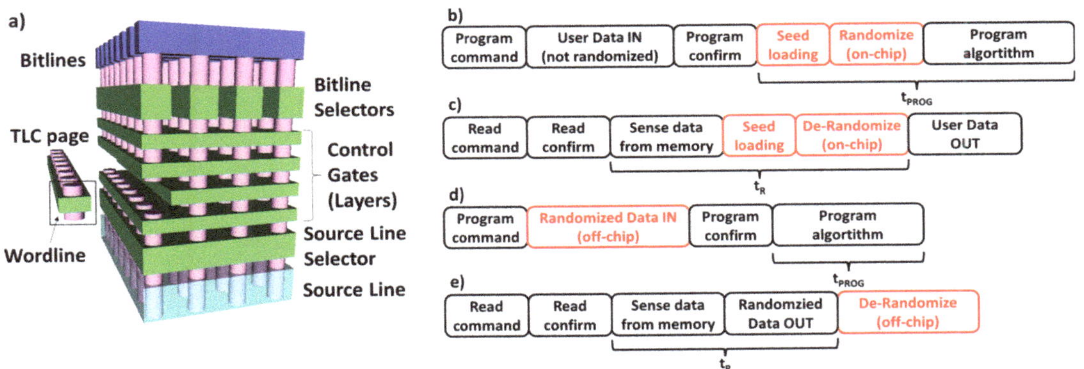

Figure 1. (**a**) The TLC 3D NAND Flash architecture. Reprinted with permission from [9] under Creative Commons License 4.0 (CC-BY). (**b**) Sequence of operations during program operation considering on-chip randomization. (**c**) Sequence of operations during read operation considering on-chip randomization. (**d**) Sequence of operations during program operation considering off-chip randomization. (**e**) Sequence of operations during read operation considering off-chip randomization.

The goal of the data randomization is that this operation will scramble the data to be sent for programming in the different memory wordlines after the data input from the host interfaces with the memory, and the de-randomization operation happens before the data output from the memory to the host starts [12,14]. Figure 1b,c shows the operation flow, considering the case of an on-chip implemented randomizer. The random seed is loaded into an internal circuit of the 3D NAND Flash memory, called a *page buffer*, via the memory data-path. Then, additional circuits take the seed and execute bit-wise XOR of the original data input from the host and random sequence in the page buffers. The program algorithm can then start. On the contrary, de-randomizing operations happen during read mode: first, the data from the memory are sensed, then the seed is loaded into a page buffer and, finally, a bit-wise XOR of sensed data and random sequence is executed making the original data available to the host. In the case of on-chip randomizers, the time for on-chip randomizing is added to the program (t_{PROG})/read (t_R) time. Off-chip randomizers can help to reduce the former times by providing a scrambled version of the data to be programmed in the memory, but in this case, it is the host that needs to take care of both randomization and de-randomization (see Figure 1d,e).

2.2. Randomizers Based on LFSRs

The principal solutions adopted for data randomization utilize a k-bit ALFSR (Autonomous Linear Feedback Shift Register), as shown in Figure 2. Feedback functions exist for any k value [18], guaranteeing that, once initialized in any state but "all zeros", the register evolves through all the $2^k - 1$ states before returning to the initial state.

The initial state is generally denoted as the register *Seed*. If the sequence generated by the ALFSR—for instance, that collected at exit Y_{k-1}—is sufficiently long, pseudorandom characteristics are guaranteed: the probability of having bits equal to 0 is 0.5, that of having any 2-bit sequence (00, 01, 10, 11) is 0.25, that of having any 3-bit sequence (000, 001, $\cdots$, 111) is 0.125 and so on.

Besides these important statistical properties, for the problem at hand it must be observed that the maximum number of consecutive zeros is equal to $k - 1$, whereas the maximum number of ones is equal to k. The former statement derives from the fact that k consecutive zeros correspond to the "all zeros" register state, which does not belong to the state diagram (otherwise the register would remain in that state because of the absence of any input). The latter derives from the fact that, when k consecutive bits equal to one are encountered, the ALFSR is in the "all one" state and, consequently, the next state must be $0111 \cdots 111$ (otherwise the ALFSR would remain in the same state, contradicting its cycling properties). The sequences of $k-1$ zeros and those of k ones occur just once in an entire 2^k-1 bit sequence.

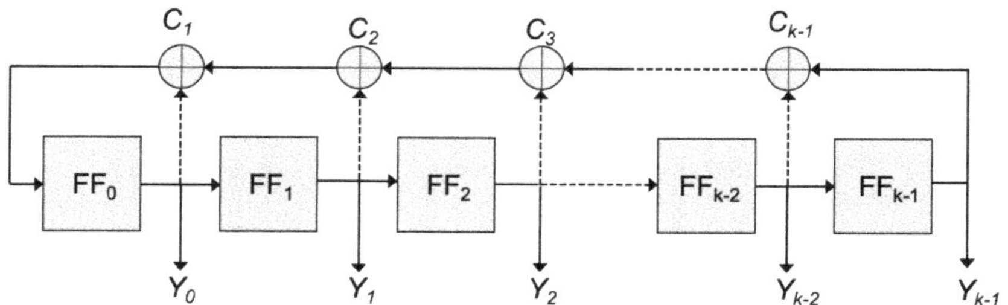

Figure 2. Schematic representation of an ALFSR. It is realized by k D-flip-flop ($FF_0 \div FF_{k-1}$) and a feedback path where some XOR C_i may be present. The feedback function depends on the presence/absence of the XOR (at least one must be present). Preset signals for register initialization are not shown. The autonomous property indicates that no input is present, so that once initialized in any state but all zeros, the cycling diagrams depend only on the feedback function.

Two possible schemes adopted to randomize the data to be stored in a memory page are shown in Figure 3. In the former, at any clock cycle c, input data $d(c)$ is XORed with the content of the last register bit $Y_{k-1}(c)$; in the latter, the ALFSR is cycled for k clock cycles, then all the register content $Y_0(c+k), \cdots, Y_{k-1}(c+k)$ is XORed with k input data ($d(c)$, $\cdots, d(c+k)$) and this procedure is repeated until all page data are randomized. If the same Seed in considered, the two schemes are fully equivalent in terms of randomization since, for the same data input sequence, they produce the same data sequence to be stored in the memory page.

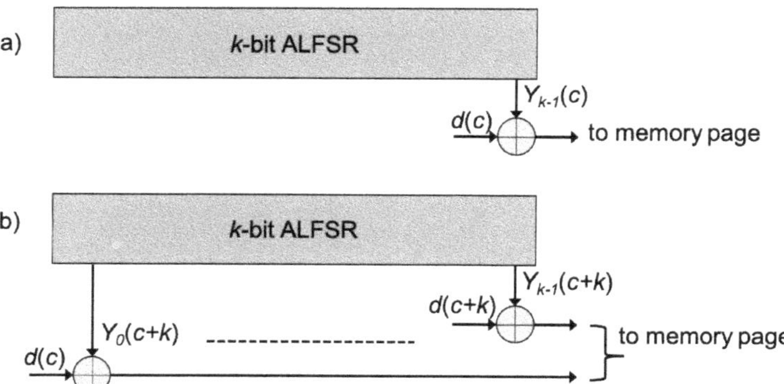

Figure 3. Possible schemes used to randomize data in a memory page. (**a**) at any clock cycle c, the input data $d(c)$ is XORed with the last register bit $Y_{k-1}(c)$; (**b**) k clock pulses are applied to the register, then the register content is XORed with k input data and the procedure is repeated until all page data are randomized.

A 3D NAND Flash memory block is constituted by N_P pages featuring N_B cells each. N_B is in the range of 4 kB $\div$ 16 kB (corresponding to $2^{15} \div 2^{17}$ cells) whereas N_P is in the range of 256 $\div$ 1024 (i.e., $2^8 \div 2^{10}$). The typical ALFSR length adopted in the randomizing schemes is $k=32$, so that a sequence of $N_L = 2^{32}-1$ can be generated by the register before returning to its initial state. Since $N_B \ll N_L$, it is clear that the statistical properties of the ALFSR are not fully exploited. A 32-bit ALFSR is generally considered a good player in data randomization.

The same ALFSR is used to randomize data for all pages in a block by changing its Seed for each page. The different Seeds can be picked from an LUT and then stored in a table, or generated internally by manipulating the page address, depending on the strategy adopted by the memory manufacturer. Unfortunately, this technique, even if providing a relatively good randomization for data stored in a page, fails at guaranteeing a good vertical data randomization along the bitline.

To illustrate the problem, Table 1 shows the sequences generated by a 4-bit ALFSR considering $N_B = N_P = 15$, obtained by randomly picking the initial Seed. As can be seen in the 4th column, all ones and all zeros are clustered, confirming that, whereas ALFSRs can be conveniently used to randomize data in the horizontal direction, no statistical predictions can be drawn when looking at a single bitline.

Table 1. Each row shows the sequence of $2^4 - 1$ bits $Y_3(c)$ generated by a 4-bit ALFSR whose initial seed is selected randomly. The 4th column enlightens the presence of long sequences of 0 or 1.

0	0	0	**1**	0	0	1	1	0	1	0	1	1	1	1
1	0	0	**1**	1	0	1	0	1	1	1	1	0	0	0
0	0	1	**1**	0	1	0	1	1	1	1	0	0	0	1
1	1	0	**1**	0	1	1	1	1	0	0	0	1	0	0
0	1	0	**1**	1	1	1	0	0	0	1	0	0	1	1
1	0	1	**1**	1	1	0	0	0	1	0	0	1	1	0
0	1	1	**1**	1	0	0	0	1	0	0	1	1	0	1
1	1	1	**1**	0	0	0	1	0	0	1	1	0	1	0
1	0	0	**0**	1	0	0	1	1	0	1	0	1	1	1
0	0	1	**0**	0	1	1	0	1	0	1	1	1	1	0
0	1	0	**0**	1	1	0	1	0	1	1	1	1	0	0
0	1	1	**0**	1	0	1	1	1	1	0	0	0	1	0
1	0	1	**0**	1	1	1	1	0	0	0	1	0	0	1
1	1	1	**0**	0	0	1	0	0	1	1	0	1	0	1
1	1	0	**0**	0	1	0	0	1	1	0	1	0	1	1

To analyze the problem in real cases, we performed simulations using a 32-bit ALFSR considering $N_P = 256$ pages, each of $N_B = 2^{17}$ cells (i.e., 16 kB). Two procedures have been selected to determine the ALFSR's Seeds: in the former, each Seed is a linear manipulation $(7 * p + 1)$ of the page address p; in the latter, each Seed is picked randomly among all the $2^{32} - 1$ possibilities.

Figure 4 shows, for the two cases, the frequency of the maximal length of zeros per bitline. Similar results are expected for ones. It must be noted that, in the first case, we observe more than 128 consecutive zeros along the bitline, thus resulting in potential reliability issues for the 3D NAND Flash memory. Figure 5 shows the distributions of the probabilities of zeros in a bitline for the two cases. Once again, we observe that the first case is critical since there are some bitlines where the number of ones and zeros is strongly unbalanced.

When the ALFSR Seeds are mathematically derived from the address page, the probability of producing a specific maximum length cluster shows a discrete spectrum: in particular, in 85 bitlines, all data are zeros. On the contrary, with the set of randomly selected Seeds used in this case, no clusters longer than the ALFSR length are found. However, it can be verified that, since the probability of zeros in a bitline shows a Gaussian-like distribution, the length of zero runs, and the probability of zeros per bitline may range over all their possible values.

However, since generally $N_P \ll 2^k - 1$, it is possible to simulate the behavior of an ALFSR considering a random Seed for each page and repeat the simulation until a set of N_P Seeds is found to guarantee, for each bitline, a number of zeros close to 50% and a predefined maximum length of clusters of consecutive ones or zeros. Then, these Seeds can be stored in an LUT integrated either on-chip or off-chip in the storage controller. Unfortunately, the quality of the set of Seeds depends on N_P and on N_B so that the procedure determining a good set of Seeds must be repeated from scratch when considering a different memory architecture.

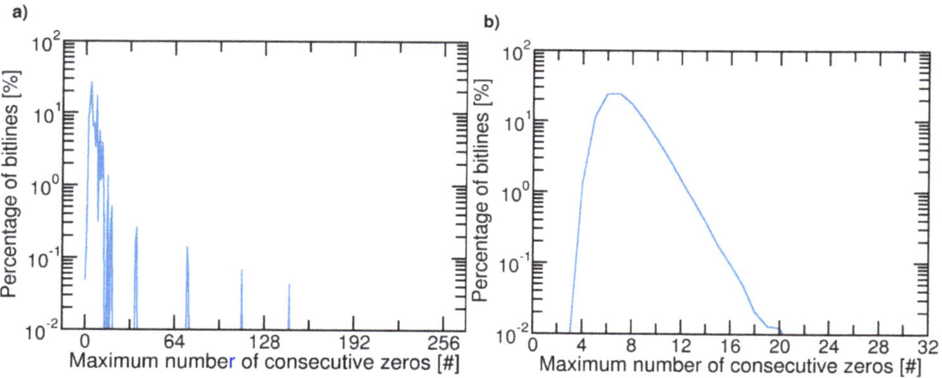

Figure 4. Percentage of bitlines as a function of the maximum number of consecutive zeros in a bitlines. Analysis has been performed considering a 32-bit ALFSR on a memory array of $N_P = 256$ pages and $N_B = 2^{17}$ cells. Case (**a**): Seeds derived from a mathematical manipulation of the page address; case (**b**): seeds generated randomly.

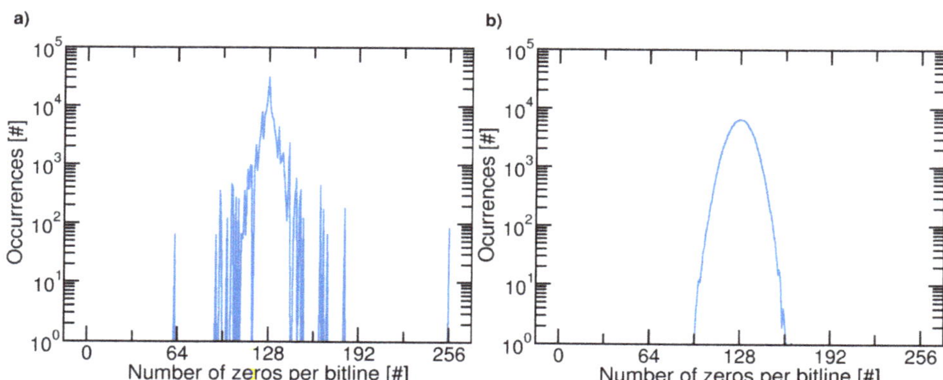

Figure 5. Distributions (occurrences) of the zeros probability in a bitline. Analysis has been performed considering a 32-bit ALFSR on a memory array of $N_P = 256$ pages and $N_B = 2^{17}$ cells. Case (**a**): Seeds derived from a mathematical manipulation of the page address; case (**b**): seeds generated randomly.

To explore the impact of random seeds selection, we randomly generated 10,000 sets, each of them containing 256 random seeds. Each set is used to feed the initial states of a 32-bit ALFSR that is used to write an array of $N_P \times B = 256 \times 2^{17}$ cells (i.e., 16 kB). For each of these 10,000 arrays, we extracted some worst case statistical parameters, namely: (i) the maximal length of consecutive zeros (ones); (ii) the maximal number of zeros (ones) in a bitline. In case (i), Figure 6a shows the number of arrays featuring a given maximal length. As can be seen, while the mean value is in the interval number of 23–24, outliers are present, featuring runs of more than 30 consecutive values. Figure 6b, instead, shows the number of arrays featuring a given maximum value of zeros. In this case, the figure also shows that outliers exist that feature a number of zeros per bitline that is larger than the average. Moreover, we must note that, as the dimension of the bitlines scales up (32 kB as shown in Figure 6), the formerly defined statistical parameters worsen.

This means that any selected set of random weights should be simulated and possibly discarded to optimize these parameters. Random selected seeds, therefore, do not represent an effective solution for the problem considered in this work. In the remainder of this work, we will propose a randomization scheme independent of the dimension of the bitlines.

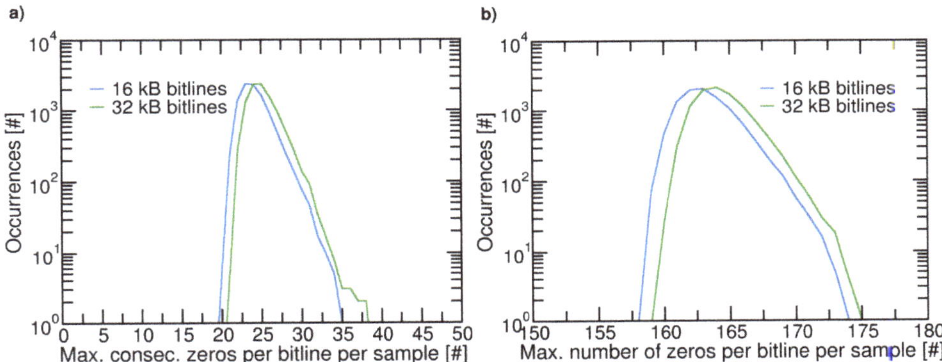

Figure 6. (**a**) Maximum consecutive number of zeros in a bitline per generated sample as a function of the bitlines' dimension; (**b**) Maximum number of zeros in a bitline per generated sample as a function of the bitlines' dimension. In these simulations we consider 256 wordlines.

3. The Proposed Solution

The solution proposed here guarantees the correct data randomization in a memory page and, at the same time, provides an upper bound for the maximum length of clusters of ones and zeros and an almost equal percentage of ones and zeros for all bitlines. The solution can be conveniently described by means of the following example: consider, for the sake of simplicity, a 4-bit ALFSR initialized with a random seed and consider the $2^4 - 1$ bit-long sequence generated as reported in the first row of Table 2. Then, consider the 2nd row as the 1st one left-shifted by one position, the 3rd row as the 2nd one left-shifted by one position and so on. We can observe that the resulting matrix is symmetrical.

By construction, it can be observed that any ith column is equal to the ith row (for instance, the 5th row and column in Table 2 are highlighted.) Therefore the statistical properties guaranteed in a $2^k - 1$ long sequence produced by a k-bit ALFSR can be found in any column: (i) the number of ones is 2^{k-1} whereas that of the zeros is $2^{k-1} - 1$; (ii) the length of the maximum sequence of ones is equal to k whereas that of the zeros is equal to $k - 1$. In addition, as already stated, the sequence of $k - 1$ zeros and that of k ones occurs just once in an entire $2^k - 1$ bit-long sequence.

Table 2. The first row shows the sequence of $2^4 - 1$ bits $Y_3(c)$ generated by a 4-bit ALFSR whose initial seed is selected randomly. Each following row is equal to the previous one left-shifted by 1 position. The resulting array is symmetrical since, by construction, any *i*th row and column are identical (for instance, the 5th row and column are enlightened).

0	0	1	0	**0**	1	1	0	1	0	1	1	1	1	0
0	1	0	0	**1**	1	0	1	0	1	1	1	1	0	0
1	0	0	1	**1**	0	1	0	1	1	1	1	0	0	0
0	0	1	1	**0**	1	0	1	1	1	1	0	0	0	1
0	**1**	**1**	**0**	**1**	**0**	**1**	**1**	**1**	**1**	**0**	**0**	**0**	**1**	**0**
1	1	0	1	**0**	1	1	1	1	0	0	0	1	0	0
1	0	1	0	**1**	1	1	1	0	0	0	1	0	0	1
0	1	0	1	**1**	1	1	0	0	0	1	0	0	1	1
1	0	1	1	**1**	1	0	0	0	1	0	0	1	1	0
0	1	1	1	**1**	0	0	0	1	0	0	1	1	0	1
1	1	1	1	**0**	0	0	1	0	0	1	1	0	1	0
1	1	1	0	**0**	0	1	0	0	1	1	0	1	0	1
1	1	0	0	**0**	1	0	0	1	1	0	1	0	1	1
1	0	0	0	**1**	0	0	1	1	0	1	0	1	1	1
0	0	0	1	**0**	0	1	1	0	1	0	1	1	1	0

Hardware Realization

The proposed solution can be easily implemented for any 3D NAND Flash memory architecture. It makes use of two *k*-bit ALFSRs, where $k = \lceil \log_2 N_P \rceil$ (i.e., $k = 8$ for $N_P = 256$; $k = 9$ for $257 < N_P \leq 512$, and so on). As shown in Figure 7, one ALFSR is used to generate the Seeds for the second ALFSR, which is used to generate the sequence randomizing the data in a page.

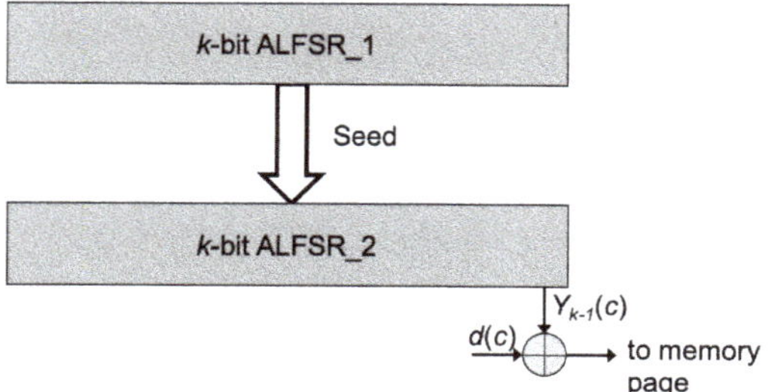

Figure 7. Proposed scheme for data randomization. For each memory page, ALFSR_1 generates the seed initializing ALFSR_2 whose content is used to randomize the data to be stored. $k = \lceil \log_2 N_P \rceil$.

The algorithm to be applied is the following:

1. After a block erase operation, initialize $ALFSR_1$ with $SEED_{IN}$. $SEED_{IN}$ can always be the same or, more conveniently to avoid a reliability degradation of 3D NAND Flash cells during endurance stress, can be generated randomly and saved in a memory location. It is mandatory to retrieve the selected $SEED_{IN}$ since it must be used to reconstruct the data sequence during a read operation;
2. Download the content of $ALFSR_1$ into $ALFSR_2$. In practice, the Seed of $ALFSR_2$ is the present state of $ALFSR_1$;

3. Program the memory page by cycling $ALFSR_2$ until the completion of the page programming while keeping $ALFSR_1$ on hold, preventing an evolution of its internal state;
4. Send a clock pulse to $ALFSR_1$ that moves to the next state;
5. Repeat steps 2 to 4 until the completion of the block programming.

By considering a $N_P = 256 \times N_B = 2^{17}$ memory block and an 8-bit ALFSR, the data provided by the proposed method consist of a sequence of 514 pseudo symmetrical arrays, each with 256 rows and 255 columns, as shown in Figure 8. In any array, the statistical properties provided by the 8-bit ALFRS are guaranteed. Since the ALFSR period is $2^k - 1$, we can avoid issues related to the logical period that are powers of two.

1	2	3		513	514
256 x 255	256 x 255	256 x 255	------------	256 x 255	256 x 255

Figure 8. When applied to a $N_P = 256 \times N_B = 2^{17}$ memory block, the data provided by an 8-bit ALFSR consist of a sequence of 256×255 arrays.

A k-bit counter is required if the memory block is not programmed sequentially page after page. The counter is preset with the index of the page to be programmed. $ALFSR_1$ is initialized with $SEED_{IN}$ as in point 1 of the described algorithm. Then a countdown starts and, with every clock pulse, $ALFSR_1$ moves to the next state. When the counter reaches the zero state, the content of $ALFSR_1$ is downloaded to $ALFSR_2$ (as in point 2 of the algorithm) since it represents the correct Seed for the page to be programmed.

When the memory block is read sequentially page after page, the same procedure used for data programming is applied. When a single page is to be read, the procedure used for non-sequential programming is applied with the counter initialized with the page address.

Data stored in the memory array can be easily reconstructed by XORing the data saved in the memory cells with the $ALFRS_2$ content, depending on the randomizing scheme (see Figure 9).

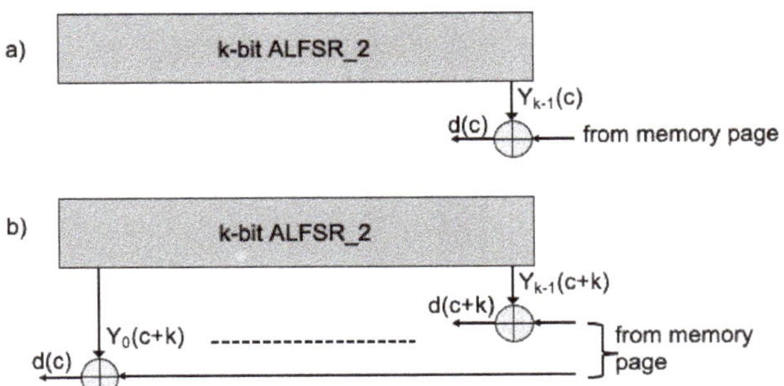

Figure 9. Possible schemes used to reconstruct data read from a memory page. (**a**) At any clock cycle c the read data is XORed with the last register bit $Y_{k-1}(c)$ to provide $d(c)$; (**b**) k clock pulses are applied to $ALFSR_2$, then the register content is XORed with k data read and the procedure is repeated until all page data are read.

4. Experimental Validation

The experimental validation of a randomization scheme requires the assessment of the memory reliability according to the data pattern written within. In this work, this activity took place by characterizing the RBER of an off-the-shelf N ($N < 100$) layers TLC 3D NAND Flash technology featuring M ($M < 16$) wordlines per layer, where its input data were supplied either by a Horizontal Centric (HC) randomizer (i.e., a randomizer that does not properly control the vertical randomization) or by our proposed method. The RBER characterization has been performed in two well-defined corners of the memory lifetime, namely after an endurance stress test (i.e., repeatedly writing and erasing the memory blocks) and after a data retention test at high temperatures. The standards adopted for endurance stress and data retention were chosen according to the JEDEC tests specifications for the 3D NAND Flash enterprise qualification procedure [19], resulting in 3000 Program/Erase cycles (that is the technology rated endurance) at a temperature of 61 °C for 500 h cycle time and a retention stress test performed on cycled devices for 3 months at 40 °C.

The experimental setup described in [20] has been exploited for both characterizations. To rule out any topological artifacts in the measurements, we disabled any error correction functionality of the chip, and we did not apply any modification to the standard working voltages of the devices and no special test modes were exploited to filter the RBER. We also ruled out the presence of on-chip randomizer circuitry that could alter the findings. The data analysis was performed on all the wordlines within a memory block considering all the TLC page types. The size of a page is 16 kB, along with the parity left for error correction code purposes, divided into 4 kB chunks, which are the minimum units read during tests by the characterization system. The testing lasted several months.

Figure 10 shows the Complementary Cumulative Distribution Function (1-CDF) as a function of the TLC page type of the RBER in a 3D NAND Flash block programmed with an HC randomizer (cases *a* and *c*) or with our proposed method (cases *b* and *d*). For the endurance stress cases we observe that an HC randomizer could dangerously induce an RBER close to the error correction capacity of many advanced schemes (we refer here to the case of a 100 b/1 kB that is a maximum allowable RBER of 1.1×10^{-2}) [21,22], whereas with our proposed method, we still have a sufficient margin with respect to that reliability limit. In the HC randomizer, we remember that there is neither a control of the patterns of ones and zeros achieved along the bitline (i.e., vertical dimension) nor an upper bound to their maximum length. This can result in some fortunate patterns (as observed in these experiments for the CSB pages), where the RBER appears as the best; however, this is at the expense of inducing the worst patterns on the other TLC pages (i.e., LSB and MSB). From the statistical standpoint, we observe an imbalanced situation where the optimal patterns are concentrated only in one TLC page type. With our proposed methodology, we guarantee a good pattern balancing among the TLC pages, thus significantly reducing the worst RBER case and homogenizing the behavior of all the 3D NAND Flash memory locations. This leads to a better control of the 3D NAND Flash reliability and in turn to a reduced system-level effort in coping with endurance and retention errors using complex error correction codes or secondary correction schemes. We also want to stress that, due to the different architectural and integration options of 3D NAND Flash technology [17], we expect that the RBER behavior may expose a different trend for memories manufactured with a different process/architecture with respect to that characterized in this work. However, our proposed randomization methodology is still foreseen to yield the same RBER improvements, while the HC randomizer will still evidence shortcomings in terms of the worst case RBER.

If we consider the retention test case, which is an additive RBER factor with respect to what we observed during the endurance test (retention tests are performed after that), we observe that all TLC pages written with an HC randomizer become uncorrectable since their RBER crosses the 1.1×10^{-2} limit, whereas in our method, only the MSB pages are above it. This suggests that, while with our method we may think to develop secondary error correction schemes [23–25] targeted only at MSB pages in retention conditions, with an HC randomizer we are forced to deal with an additional effort to recover the data every time we access the 3D NAND Flash after a data retention stress.

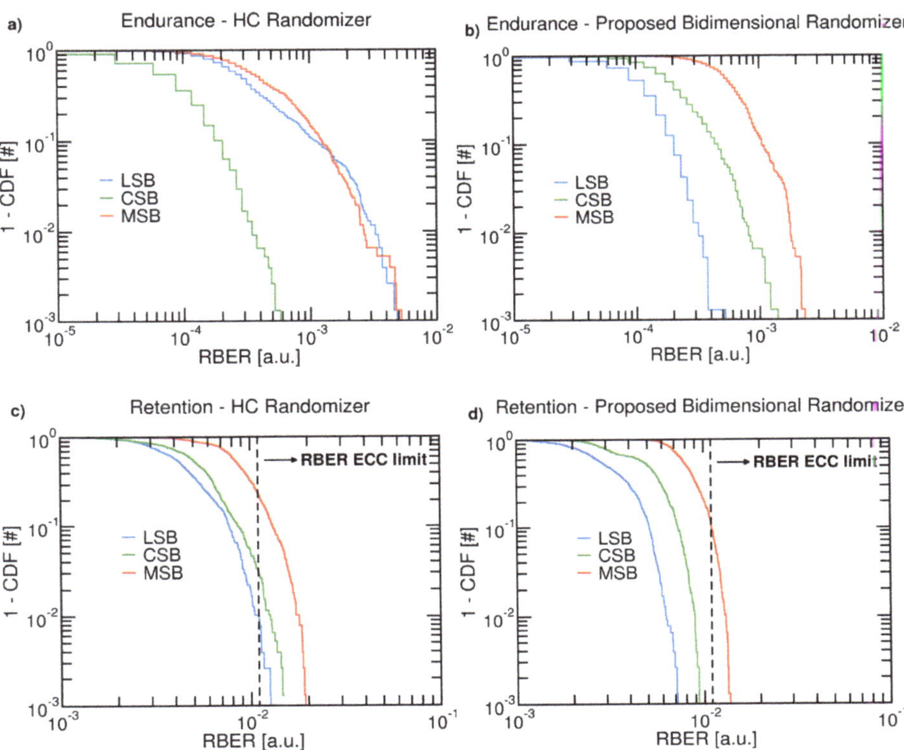

Figure 10. Complementary Cumulative Distribution Function (1-CDF) of the RBER in a 3D NAND Flash memory for Endurance and Retention working corners when the input data come from an HC randomizer (**a**–**c**) or from our proposed method (**b**–**d**).

Finally, Figure 11 shows the results of a topological characterization of the RBER in a 3D NAND Flash block. As can be seen, there are specific areas (i.e., layers and wordlines) for which the use of a good quality randomizer could help in terms of improving the reliability during both endurance and retention working conditions. We must note that, in the HC randomizer, the presence of uncontrolled sequences of consecutive ones and zeros, coupled with the non-perfect randomization along the vertical dimension of the memory, severely affects the sensing operation of the data with the consequent burden on the RBER. In 3D NAND Flash memory architectures (please refer to Figure 1a), the layers are the contacts stacked along the vertical dimension (let us refer to it as the y-axis) also referred to as the control gates to which the voltages for programming and reading the memory are applied. For each layer, there are several wordlines associated and connected in the direction of the z-axis, so that a single layer (control gate) can drive the signal in parallel on multiple wordlines. The bitlines are in the x-axis direction. Let us assume a total of five wordlines per layer. Since we have a TLC storage paradigm, we will come up with five wordlines associated with LSB, five wordlines for CSB, and five wordlines for MSB. That is why we have well-defined "stripes" in the plots of Figure 11. The higher wordline indexes are associated with MSB pages and the lower indexes to LSB. In this case, Figure 11 reflects the same behavior as observed in Figure 10. Concerning the variability of RBER characteristics along the layers and the wordlines, we must note that the 3D NAND Flash manufacturing process is not easy to control, so there is a well-known sensitivity of the RBER's layers that depends exactly on the peculiar processing steps of the memory devices, which, unfortunately, are not disclosed to us.

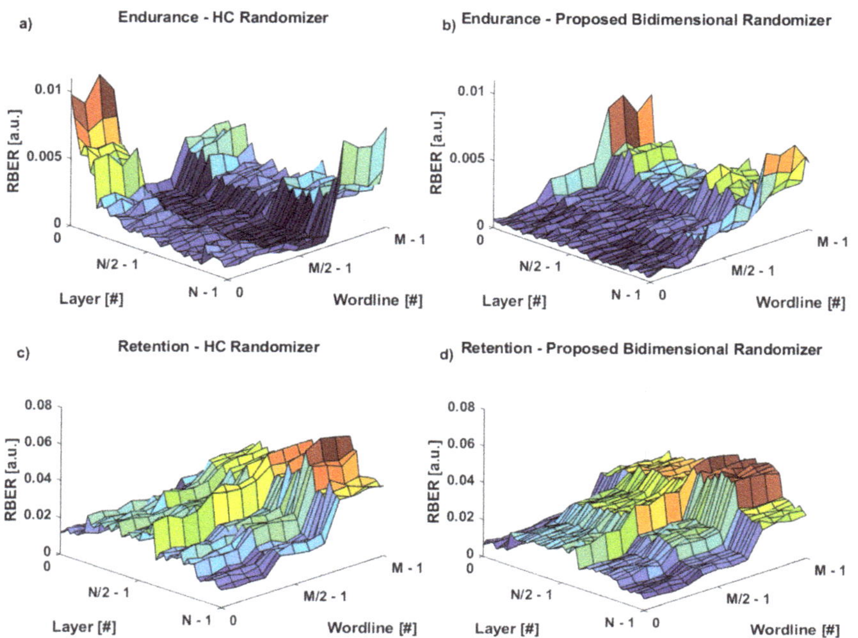

Figure 11. Topological characterization of the RBER in a 3D NAND Flash memory for Endurance and Retention working corners when the input data come from an HC randomizer (**a**,**c**) or from our proposed method (**b**,**d**).

5. Conclusions

In this work, we proposed a randomization scheme for 3D NAND Flash memory technology that allows a good degree of randomization in both memory dimensions (i.e., wordline and bitline) without requiring a complex implementation methodology while relying only on a proper arrangement of LFSRs circuits. The simulation results show that our methodology imposes a guard band on the maximum number of consecutive ones and zeros along the bitline dimensions (no more than 25) to keep the read failure probability during the data readout phase under control.

Further, we demonstrate by construction that our randomization scheme has better control of the number of zeros or ones along the bitline, proving a good balancing of the write data to the memory, thus representing an optimal case for reliability.

Finally, we experimentally validated our proposed methodology on an off-the-shelf TLC 3D NAND Flash memory chip, showing that, under JEDEC-style endurance and data retention stress, we can achieve RBER for LSB and CSB pages that is always below the correction limit imposed by a 100 b/1 kB Error Correction Code. Our randomization methodology can therefore be exploited by storage system designers to keep the memory reliability under control.

Author Contributions: The individual contributions to this paper are the following: conceptualization, M.F., P.O. and R.M.; methodology, M.F., A.M. and P.O.; software, M.F.; validation, A.M. and R.M.; electrical measurements and resources, A.M.; investigation, C.Z., P.O. and M.F.; data curation, C.Z. and R.M.; writing–original draft preparation, M.F. and C.Z.; writing–review and editing, A.M., R.M. and P.O.; visualization, C.Z.; supervision, R.M. and P.O. All authors have read and agreed to the published version of the manuscript.

Funding: This research no external funding.

Conflicts of Interest: The authors declare no conflict of interest.

Abbreviations

The following abbreviations are used in this manuscript:

MMC	MultiMedia Card
SSD	Solid State Drive
RBER	Raw Bit Error Rate
LFSR	Linear Feedback Shift Register
LUT	Look Up Table

References

1. Li, Y. 3D NAND Memory and Its Application in Solid-State Drives: Architecture, Reliability, Flash Management Techniques, and Current Trends. *IEEE Solid State Circuits Mag.* **2020**, *12*, 56–65. [CrossRef]
2. Micheloni, R.; Aritome, S.; Crippa, L. Array Architectures for 3-D NAND Flash Memories. *Proc. IEEE* **2017**, *105*, 1634–1649. [CrossRef]
3. Righetti, N.; Puzzilli, G. 2D vs 3D NAND technology: Reliability benchmark. In Proceedings of the IEEE International Integrated Reliability Workshop (IIRW), South Lake Tahoe, CA, USA, 8–12 October 2017; pp. 1–6. [CrossRef]
4. Lin, W.; Hsu, Y.C.; Kuo, T.H.; Yang, Y.S.; Chen, S.W.; Tsao, C.W.; Liu, A.C.; Ou, L.Y.; Wang, T.C.; Yen, S.W.; et al. 3X endurance enhancement by advanced signal processor for 3D NAND flash memory. In Proceedings of the IEEE Information Theory Workshop (ITW), Kaohsiung, Taiwan, 6–10 November 2017; pp. 201–203. [CrossRef]
5. Mizoguchi, K.; Takahashi, T.; Aritome, S.; Takeuchi, K. Data-Retention Characteristics Comparison of 2D and 3D TLC NAND Flash Memories. In Proceedings of the IEEE International Memory Workshop (IMW), Monterey, CA, USA, 14–17 May 2017; pp. 1–4. [CrossRef]
6. Zambelli, C.; Crippa, L.; Micheloni, R.; Olivo, P. Cross-Temperature Effects of Program and Read Operations in 2D and 3D NAND Flash Memories. In Proceedings of the IEEE International Integrated Reliability Workshop (IIRW), South Lake Tahoe, CA, USA, 7–11 October 2018; pp. 1–4. [CrossRef]
7. Cai, Y.; Luo, Y.; Ghose, S.; Mutlu, O. Read Disturb Errors in MLC NAND Flash Memory: Characterization, Mitigation, and Recovery. In Proceedings of the IEEE/IFIP International Conference on Dependable Systems and Networks, Rio de Janeiro, Brazil, 22–25 June 2015; pp. 438–449.
8. Resnati, D.; Goda, A.; Nicosia, G.; Miccoli, C.; Spinelli, A.S.; Compagnoni, C.M. Temperature Effects in NAND Flash Memories: A Comparison Between 2-D and 3-D Arrays. *IEEE Electron Device Lett.* **2017**, *38*, 461–464. [CrossRef]
9. Zambelli, C.; Micheloni, R.; Scommegna, S.; Olivo, P. First Evidence of Temporary Read Errors in TLC 3D-NAND Flash Memories Exiting From an Idle State. *IEEE J. Electron Devices Soc.* **2020**, *8*, 99–104. [CrossRef]
10. Zuolo, L.; Zambelli, C.; Micheloni, R.; Olivo, P. Solid-State Drives: Memory Driven Design Methodologies for Optimal Performance. *Proc. IEEE* **2017**, *105*, 1589–1608. [CrossRef]
11. Viterbi, A. Approaching the Shannon Limit: Theorist's Dream and Practitioner's Challenge. In *Mobile and Personal Satellite Communications 2*; Vatalaro, F., Ananasso, F., Eds.; Springer: London, UK, 1996; pp. 1–11.
12. Yoon, S. Nonvolatile Memory Device and Data Randomizing Method Thereof. U.S. Patent 20100259983A1, 14 October 2010. Available online: https://patents.google.com/patent/US20100259983A1/en (accessed on 23 May 2021).
13. Cernea, R.A. Highly Compact Non-Volatile Memory and Method Thereof. U.S. Patent 2004/0060031A1, 25 March 2004. Available online: https://patents.google.com/patent/US20040060031A1/en (accessed on 21 May 2021).
14. Kim, C.; Ryu, J.; Lee, T.; Kim, H.; Lim, J.; Jeong, J.; Seo, S.; Jeon, H.; Kim, B.; Lee, I.; et al. A 21 nm High Performance 64 Gb MLC NAND Flash Memory With 400 MB/s Asynchronous Toggle DDR Interface. *IEEE J. Solid State Circ.* **2012**, *47*, 981–989. [CrossRef]
15. Cha, J.; Kang, S. Data Randomization Scheme for Endurance Enhancement and Interference Mitigation of Multilevel Flash Memory Devices. *ETRI J.* **2013**, *35*, 166–169. [CrossRef]
16. Atsumi, T.; Kurosawa, Y. Nonvolatile Memory Device and Data Randomizing Method Thereof. U.S. Patent 20170160939A1, 8 June 2017. Available online: https://patents.google.com/patent/US20170160939A1/en (accessed on 23 May 2021).
17. Micheloni, R.; Crippa, L.; Zambelli, C.; Olivo, P. Architectural and Integration Options for 3D NAND Flash Memories. *Computers* **2017**, *6*, 27. [CrossRef]
18. Golomb, S.W. Shift Register Sequences—A Retrospective Account. In Proceedings of the 4th International Conference on Sequences and Their Applications, Beijing, China, 24–28 September 2006; pp. 1–4. [CrossRef]
19. JEDEC. *JESD22-A117 Electrically Erasable Programmable ROM (EEPROM) Program / Erase Endurance and Data Retention Stress Test*; JEDEC: Arlington, VA, USA, 2018.
20. Zambelli, C.; Micheloni, R.; Crippa, L.; Zuolo, L.; Olivo, P. Impact of the NAND Flash Power Supply on Solid State Drives Reliability and Performance. *IEEE Trans. Device Mater. Reliab.* **2018**, *18*, 247–255. [CrossRef]
21. Li, Q.; Shi, L.; Cui, Y.; Xue, C.J. Exploiting Asymmetric Errors for LDPC Decoding Optimization on 3D NAND Flash Memory. *IEEE Trans. Comput.* **2020**, *69*, 475–488. [CrossRef]
22. Zhang, M.; Wu, F.; Yu, Q.; Liu, W.; Cui, L.; Zhao, Y.; Xie, C. BeLDPC: Bit Errors Aware Adaptive Rate LDPC Codes for 3D TLC NAND Flash Memory. In Proceedings of the Design, Automation and Test in Europe (DATE), Grenoble, France, 9–13 March 2020; pp. 302–305. [CrossRef]

23. Wang, J.; Courtade, T.; Shankar, H.; Wesel, R.D. Soft Information for LDPC Decoding in Flash: Mutual-Information Optimized Quantization. In Proceedings of the IEEE Global Telecommunications Conference, Houston, TX, USA, 5–9 December 2011; pp. 1–6. [CrossRef]
24. Im, S.; Shin, D. Flash-Aware RAID Techniques for Dependable and High-Performance Flash Memory SSD. *IEEE Trans. Comput.* **2011**, *60*, 80–92. [CrossRef]
25. Kim, J.; Lee, E.; Choi, J.; Lee, D.; Noh, S.H. Chip-Level RAID with Flexible Stripe Size and Parity Placement for Enhanced SSD Reliability. *IEEE Trans. Comput.* **2016**, *65*, 1116–1130. [CrossRef]

 micromachines 

Article

Observation and Optimization on Garbage Collection of Flash Memories: The View in Performance Cliff

Yajuan Du [1,*], Wei Liu [1], Yuan Gao [1] and Rachata Ausavarungnirun [2,*]

1 School of Computer Science and Technology, Wuhan University of Technology, Wuhan 430070, China; hellovemail@163.com (W.L.); gaoyuan0107@foxmail.com (Y.G.)
2 Sirindhorn International Thai-German Graduate School of Engineering, King Mongkut's University of Technology North Bangkok (KMUTNB), Bangkok 10800, Thailand
* Correspondence: dyj@whut.edu.cn (Y.D.); r.ausavarungnirun@gmail.com (R.A.)

Abstract: The recent development of 3D flash memories has promoted the widespread application of SSDs in modern storage systems by providing large storage capacity and low cost. Garbage collection (GC) as a time-consuming but necessary operation in flash memories largely affects the performance. In this paper, we perform a comprehensive experimental study on how garbage collection impacts the performance of flash-based SSDs, in the view of performance cliff that closely relates to Quality of Service (QoS). According to the study results using real-world workloads, we first observe that GC occasionally causes response time spikes, which we call the performance cliff problem. Then, we find that 3D SSDs exacerbate the situation by inducing a much higher number of page migrations during GC. To relieve the performance cliff problem, we propose PreGC to assist normal GC. The key idea is to distribute the page migrations into the period before normal GC, thus leading to a reduction in page migrations during the GC period. Comprehensive experiments with real-world workloads have been performed on the SSDsim simulator. Experimental results show that PreGC can efficiently relieve the performance cliff by reducing the tail latency from the 90th to 99.99th percentiles while inducing a little extra write amplification.

Keywords: solid-state drives; 3D flash memory; performance cliff; tail latency; garbage collection

Citation: Du, Y.; Liu, W.; Gao, Y.; Ausavarungnirun, R. Observation and Optimization on Garbage Collection of Flash Memories: The View in Performance Cliff. *Micromachines* **2021**, *12*, 846. https://doi.org/10.3390/mi12070846

Academic Editors: Cristian Zambelli and Rino Micheloni

Received: 6 June 2021
Accepted: 27 June 2021
Published: 20 July 2021

1. Introduction

Due to shock-resistance, high access speed, low energy consumption, and increased capacity, Solid-State Drives (SSDs) [1–3] gradually gain popularity as the main storage device or data buffer on modern big data or AI applications [4–8]. The development of new flash memories such as 3D-stacked charge-trap (CT)-based ones largely benefits the storage density of modern SSDs. Meanwhile, they show some new physical characteristics, e.g., the increased block size and layer speed variation, the effect of which on performance have not been fully investigated [9].

Garbage collection (GC) is responsible for reclaiming blocks with a large proportion of invalid pages. A GC operation consists of two main phases: valid page migration and block erase. GC often has a great impact on system performance. Paik et al. [10] and Wu et al. [11] considered avoiding GC blocking on read requests by directly delaying GC or by exploiting the data redundancy of multiple SSD arrays. Chen et al. [12] proposed an erase efficiency boosting strategy to reduce block erase latency by exploiting the multi-block erase characteristic of 3D CT-based SSDs. ShadowGC [13] was designed to hide GC latency by exploiting the host-side and device-side write buffers. Yan et al. [14] proposed a Tiny-Tail Flash to hide GC latency in paralleled and redundant SSD structures. Choi et al. [15] and Guo et al. [16] proposed scheduling I/O requests and GC operations together by considering the paralleled structure of SSDs. Shahidi et al. [17] combined a cache management policy with GC and proposed CachedGC to postpone writing back valid pages during the GC.

In this paper, we perform a comprehensive experimental study on how garbage collection affects the system performance of SSDs in the view of performance cliff that closely relates to tail latency and affects Quality of Service (QoS). According to preliminary study results, we first observe that SSD response time shows occasional spikes. By comparing with 2D SSDs, these spikes in 3D SSDs have much higher values and occur more frequently, which makes the performance situation worse. We call this phenomenon of response time spikes the problem of "performance cliff". This directly induces the sharp increase of tail latency that is often used as the evaluation of Quality of Service (QoS) by the industry.

In order to study the cause of performance cliff, we collect some experimental results about garbage collection and obtain two extra observations. On the one hand, the number of page migrations during GC sharply increases, especially in 3D SSDs. On the other hand, page migration latency takes up the majority of GC latency while block erase only takes up a small proportion.

According to the above observations, we propose a GC-assisting method, called PreGC to mitigate the GC latency and to optimize tail latency. The key idea of PreGC is to migrate part of valid pages in advance of normal GC, which can distribute heavy page migrations into the other system time. The challenge to implement PreGC is to decide when and how many pages to migrate. PreGC is invoked near by the normal GC and migrates valid pages during system idle time in a fine-grained incremental way. In this way, it can mitigate unnecessary migrations that overlap with page updates between PreGC and normal GC and reduces the effect of pre-migrations on normal requests.

In order to evaluate the proposed PreGC, we perform a comprehensive experiment on the SSDsim simulator with real-world workloads. From the experimental results, we show that PreGC is effective in reducing page migrations and optimizing system performance with reduced 90th to 99.99th percentile tail latencies.

The contributions of this paper are listed as follows:

- We perform a preliminary experimental study on the response time and tail latency of SSDs and observe the performance cliff problem.
- We uncover that the main cause of the performance cliff problem is the significantly increased latency of garbage collection. These increased latency are mostly caused by the increased number of page migrations in 3D SSDs.
- According to the above observations, we propose a GC-assisting method called PreGC to relieve the performance cliff. By pre-migrating a part of valid pages ahead of normal GC time, page migration latency can be distributed into other system time and thus can be largely reduced during GC period.
- We evaluate the proposed PreGC with real-world workloads on the SSDsim simulator. The results show that performance cliff can be significantly relieved by lowering down the tail latency.

The rest of this paper is organized as follows. Section 2 presents the basics of 3D SSDs and studies related works to SSD performance optimization. Section 3 illustrates the details of our preliminary study experiment and observations on 2D SSDs and 3D SSDs. Section 4 describes the detailed designs of PreGC. The experiment setup and evaluation results of PreGC are presented in Section 5. Section 6 concludes this paper.

2. Background and Related Works

This section first introduces the basic structure of 3D SSDs, in which the large block problem is mentioned. Then, we illustrate the mechanism of garbage collection. At last, layer speed variations are illustrated to show the uneven data hotness problem in 3D charge-trap (CT) SSDs.

2.1. Basics of 3D SSDs

SSDs are composed of a controller and flash arrays. The controller is responsible for organizing data access on flash arrays and for effectively using a flash. For example, the flash translation layer is used to manage the mapping between physical addresses and logical addresses. The garbage collection mechanism cleans invalid data blocks to overcome the out-of-place nature of flash memory. Moreover, error correction and wear leveling are designed to make data reliable and to even cause wear on flash blocks.

The flash arrays in 3D SSDs are composed of 3D flash memory, which greatly increases the capacity of SSDs by vertically stacking multiple layers. Figure 1 illustrates the physical organization of flash cells in 3D flash memory. The control gates of the cells belonging to the same layer are connected together to form a wordline. All cells with the same bitline across multiple layers form a block. It can be found that the block size would be sharply increased because of the layer stacking, compared with 2D flash memory. This induces the big block problem that has been widely studied in existing works [18,19]. When the block size is larger, the block erase time and migrated page numbers would be prolonged, which induces worse garbage collection performance as well as long tail latency.

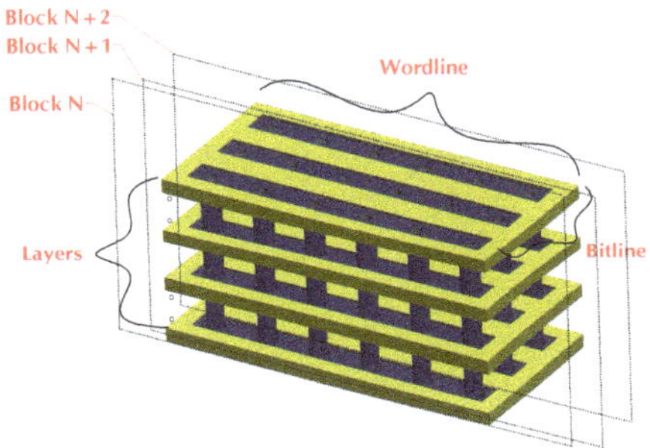

Figure 1. The layer-stacked structure of 3D flash memory.

Due to the out-of-place update feature of flash memory, a lot of invalid data would be generated after SSD has been used for a while. Garbage collection is used to reuse the space occupied by these invalid data. The granularity to perform GC is a block, but the basic unit of read and write is page. The process of GC is mainly divided into two stages: valid data migration and block erase. After a victim block is selected, valid pages are first migrated into another block. After all valid pages are migrated, block would be erased to be a free block again. Thus, the latency of GC is decided not only by block erase but also by page migrations.

2.2. Layer Speed Variations

This part introduces the charge trap (CT)-based flash memory, a special type of 3D flash memory widely used in 3D SSDs, which utilizes an effective way to construct a vertical flash structure. There are multiple gate stack layers and vertical cylinder channels in 3D CT flash [20,21], as shown in Figure 2. A special chemical liquid is used to erode the stacked layers. The physical properties of this liquid cause the upper layer to have a larger opening than lower layers, which leads to asymmetric feature process size across the stacked layers. The electric field strength of each layer is different, and for the larger opening, the electric field strength would be high, which induces a slower access speed. Thus, access speed on lower layers is faster than that on upper layers. This phenomenon is called the layer speed variations.

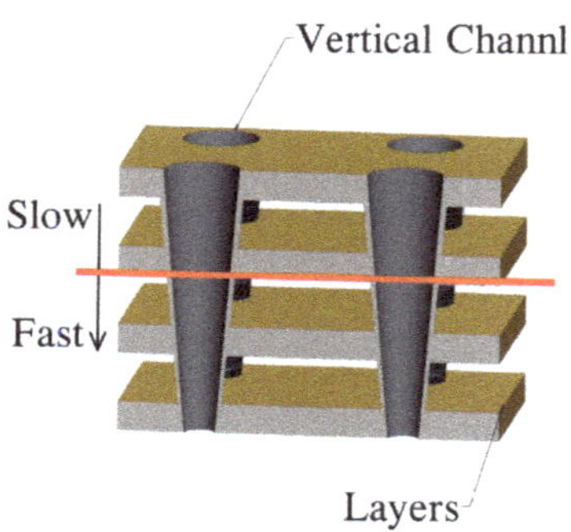

Figure 2. Three-dimensional CT-based flash.

2.3. Related Works

This paper focuses on optimizing the performance of 3D SSDs in the view of garbage collection. As previous works related to garbage collection schemes have been discussed in Section 1, this section investigates existing works that optimize 3D SSD performance, most of which study or exploit the special characteristics of 3D layer-stacked structures. In detail, these characteristics can be divided into two types: the logic in programming and reads, and the physical feature of layer-to-layer structures such as process variations. We discuss these existing works as follows.

By utilizing the logic in programming and reads, several works have been proposed. Wu et al. [22] proposed a new data allocation policy to exploit the special one-shot programming scheme in CT-based 3D flash memories. Logically sequential data are re-distributed into different parallel units to enhance read parallelism. Shihab et al. [23] relieved the fast voltage drift problem of 3D flash by applying an elastic read reference scheme (ERR) to reduce read errors, which can decrease read latency with advanced ECC codes. ApproxFTL [24] considers storing data by reducing the maximal threshold voltage and by applying an approximate write operation to store error-resilient data Pletka et al. [25] studied the shifts of threshold voltage distributions in 3D flash memory and proposed a new framework to manage 3D TLC flash errors for high SSD performance and lifetime. Ho et al. [26] proposed a one-shot program design to accelerate programming speed of 3D flash memories and to reduce data error rates. Zhang et al. [27] considered to improve the read performance of 3D SSDs in the view of ECC efficiency and proposed a RBER aware multi-sensing scheme to decrease the number of read thresholds.

By exploiting the physical feature of layer-to-layer structures, other works have been proposed. Chen et al. [28] exploited the asymmetric speed feature across layers of CT-based 3D flash and proposed a progressive scheme to boost access performance. Chen et al. [12] optimized the garbage collection performance in the view of block erase efficiency and proposed a multi-block erase strategy. Xiong et al. [29] and Wu et al. [30] studied the characteristics and challenges of 3D flash memories with the floating-gate (FG) type and the charge-trap (CT) type, respectively. Hung et al. [31] studied the cross-layer process variation problems of 3D vertical-gate flash and proposed three layer-aware program-and-read scheme to reduce P/E cycle numbers and to improve read performance. Liu et al. [32] proposed a new read operation called "single-operation-multiple-location" for small reads to enhance the chip-level parallelism of 3D NAND SSDs. Wang et al. [33] proposed a reliability management method, named as P-Alloc to tolerate process variation of 3D CT flash. As our proposed PreGC method considers the effect of layer-to-layer speed variations on GC performance, it belongs to this category. In addition, we are the first work to uncover the root cause of the performance cliff problem in 3D SSDs.

Different from the above method of hiding the necessary latency or a method of improving the long tail latency by reducing the frequency of GC blocking I/O such as GFTL [34], which provides deterministic service guarantees by leveraging the request intervals to perform partial GC, and AGC+DGC [35], which significantly reduces GC overhead to provide stable SSD performance by scheduling GC operations from busy to idle periods, our work assists GC in improving performance by reducing the time that GC blocks I/O in a novel way and is orthogonal with these works.

3. Preliminary Study

This section presents our preliminary study on 3D SSD performance based on the two problems of big block size and data unevenness. First, we introduce the experimental setup of this study, including 3D SSD configurations and workloads. Then, three observations from the studied results are explained in details. At last, through analysis and comparison, it is concluded that sharply increased page migrations during GC are the main cause of severe performance cliffs in 3D SSDs.

3.1. Experiment Setup

We used SSDsim to simulate 2D SSDs, and some of its components were modified to simulate 3D SSDs by adding layer information for data. The parameter configurations for 2D and 3D SSDs are shown in Table 1. The variation of the layer difference was simulated as the fastest layer speed was twice the speed of the slowest layer, and the middle layer gradually increased in speed. The number of pages per block in 3D SSDs were set as the double of that in 2D SSDs and the other parameters were set as the same value to reflect the big block size problem.

Table 1. Parameter configurations of 2D SSDs and 3D SSDs.

Parameter vs. Type	2D SSDs	3D SSDs
Overall capacity	16 G	16 G
Page size	4 k	16 k
Page number per block	64	128
Page read latency (μs)	20	90
Page write latency (μs)	200	1100
Block erase latency (μs)	1500	10,000
GC Threshold (μs)	10%	10%
Over-provisioning (μs)	20%	20%

Six real-world workloads [36] were chosen and are shown in Table 2, in which usr0 is a user workload and the remaining five are the workloads from the server. As the read/write request ratios and average request interval time of these workload are different, the experiment results are more representative for various applications.

Table 2. Statistics of six real-world workloads.

Trace vs. Stat	Reads	Writes	Read Ratio	Averge Interval Time (ns)
usr0	903,491	1,333,345	40%	27,037,999,239
src0	176,729	1,381,085	11%	44,862,657
ts0	316,689	1,484,799	18%	38,800,309
rsrch0	133,625	1,300,030	9%	42,185,614
fiu_web	78,613	5,604,382	1%	105,356,789
mds1	133,625	1,300,030	93%	36,646,192

3.2. Observations on SSD Performance

Based on these settings, the SSD performance cliff by GC was first observed by analyzing request response time series. Then, in order to find the reason behind this phenomenon, extra two experimental results including migrated page numbers and latency distribution in the GC period were then shown and analyzed.

3.2.1. The Problem of Performance Cliff

A main indicator for SSD performance is its response time, which is the latency in processing read and write requests. Request response time during a period of about two milion requests in the workload hm0 was collected and shown in Figure 3 and Figure 4. It can be seen that response time peaks occasionally appear both in 2D and 3D SSDs, which we call the performance cliff problem. In addition, through the comparison of two figures, it can be seen that the performance cliff of 3D SSDs is far more serious than that of 2D SSDs. We further study this phenomenon in the following sections.

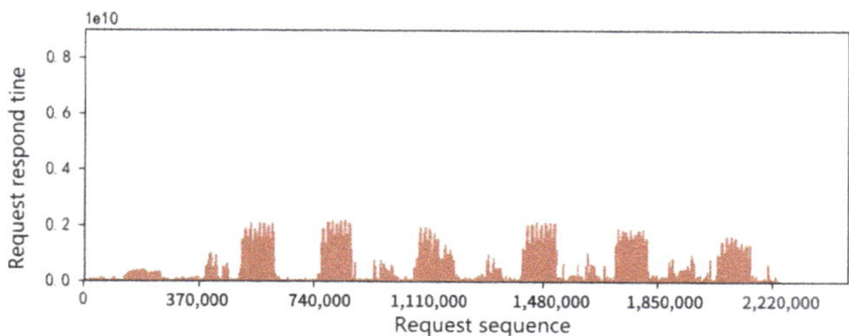

Figure 3. Request response time distribution in 2D SSDs.

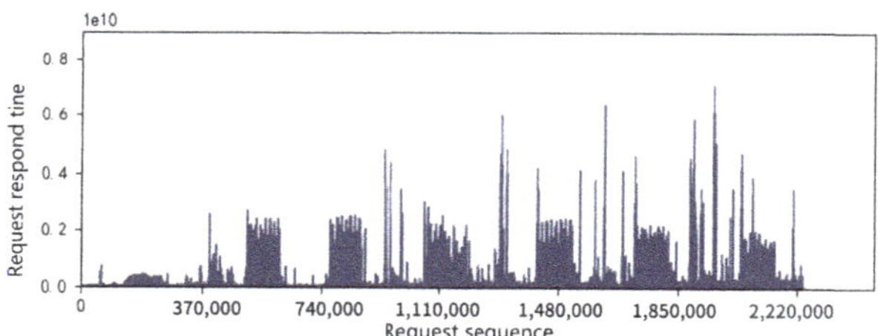

Figure 4. Request response time distribution in 3D SSDs.The performance cliff phenomenon of 3D SSDs is much more serious than that of 2D, which is manifested in a sudden high latency as shown in the figure.

3.2.2. The Number of Page Migrations

As GC performance in 3D SSDS is affected by the big block problem, which would induce increased page migrations, we collected page migrations numbers of each GC in workload hm0, as shown in Figure 5. From the figure, we can see that the number of valid pages to be migrated in GC of 3D SSDs has a sharp increase with respect to 2D SSDs when serving the same traces. Additionally, when the GC number increased, the page migration difference between two SSDs increases greatly. These results show that 3D SSDs migrated more pages as a larger block size was used, latency induced by these migrations would also be high, as shown in the next study.

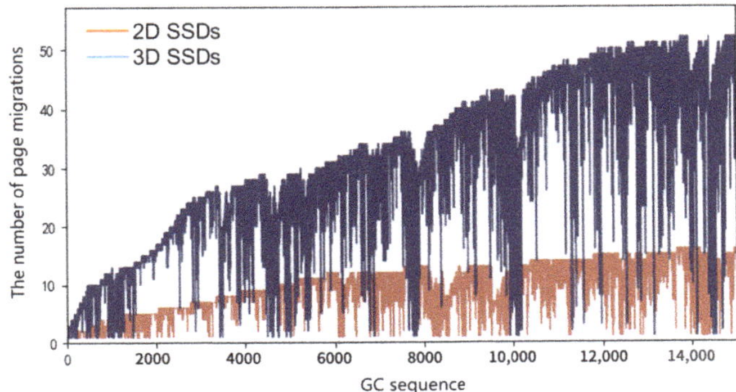

Figure 5. The number of page migrations in garbage collection. The abscissa in the figure is the serial number of GC, and the ordinate represents the number of page migrations in the current GC. The number of GC page migrations is significantly higher in 3D SSDs (**blue broken line**) than in 2D SSDs (**red broken line**).

3.2.3. Latency Distribution in GC

As illustrated in Section 2.1, the latency caused by GC is mainly composed of the latency of page migrations and block erase. This section analyzes the latency distribution of these two stages among the overall GC latency, as shown in Table 3. In this table, not only the latency distribution in GC but also the times of page migrations on block erase are presented. It can be seen from the results that the proportion of page migrations in 3D SSDs significantly increases when compared with that in 2D SSDs. For the workload src0, the latency of page migrations can reach up to 11.45 times that of block erase in 3D SSDs, while this value only reaches to 5.23 in 2D SSDs.

Table 3. Distribution of GC latency on page migration and block erase.

Ratio vs. Workload	usr0	src0	ts0	rsrch0	wdev0
2D erase	44%	16%	40%	89%	34%
2D migration	56%	84%	60%	11%	66%
migration/erase	1.28	5.25	1.52	0.12	1.95
3D erase	9%	8%	20%	23%	26%
3D migration	91%	92%	80%	77%	74%
migration/erase	9.98	11.45	4.10	3.29	2.85

As the block erase time for both SSDs is similar because of the technology development of 3D flash memory, the latency of page migrations is the main cause of high GC latency. Therefore, the server performance cliff problem of 3D SSDs uncovered above is mainly caused by the sharply increased number of page migrations. According to this conclusion, this paper proposes a reduction in page migrations for 3D SSDs by pre-migrating valid pages near the time when GC is invoked. Next, the detailed design of our method would be presented.

4. The PreGC Method

This section introduces our proposed PreGC method from three aspects: overview, workflow, and cooperation with normal GC. First, the architectural overview of PreGC is presented. Then, the workflow of PreGC is illustrated to show when to trigger PreGC, how to perform page migrations in PreGC, and when to stop these migrations. Lastly, how PreGC can assist normal GC for performance cliff reduction is shown.

4.1. Overview

The overview of 3D SSDs with PreGC is shown in Figure 6, in which the SSD controller acts as the medium for communication between the host and the storage. The SSD controller mainly includes some components such as host interface, RAM, processor, and FTL. The host interface is used to interact with the host, the RAM is used to store mapping tables between physical addresses, and logical addresses are used to facilitate data read and placement. The processor manages the request flows and performs some basic computations for SSD control algorithms.

As PreGC is a method of performing partial page migrations ahead of normal GC time, it has to work together with existing GC methods. PreGC mainly contains two components to judge when to invoke and stop the pre-migration operations: invoking and stopping. Briefly speaking, the invoking condition depends on the ratio of free blocks, which is similar to that in normal GC. However, in order to make a balance between write amplification and GC page migration reduction, the threshold ratio for invoking PreGC should be deliberately designed. The stopping condition of PreGC depends on how many valid pages exist in the victim block. As there is no need to migrate all valid pages, which may make normal GC ahead of its original, the threshold ratio is set to a value a little below the invoking threshold of normal GC. Details of the workflow to use PreGC within the right module of Figure 6 are presented next.

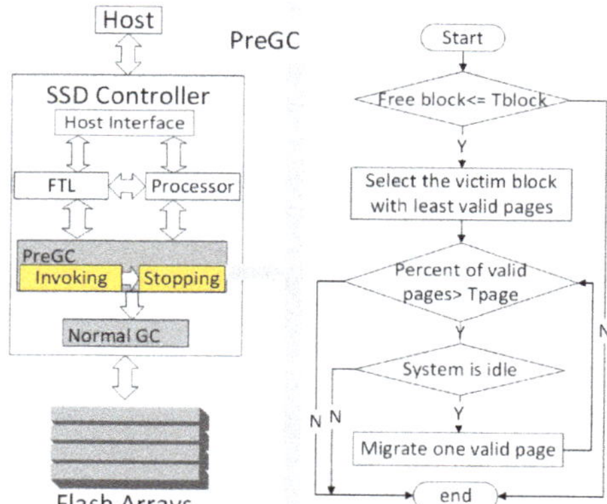

Figure 6. Overview of PreGC in 3D SSD controller. The Pregc mechanism is located in the SSD controller and works with the FTL, processor, etc., including the invoking module and the stopping module; the workflow of Pregc is shown on the right.

4.2. Workflow of PreGC

In order to better describe the specific implementation process of PreGC, a workflow chart is presented in the right part of Figure 6. It mainly involves three judgements, the invoking and stopping conditions of page pre-migration operations, and the current system status. Two threshold parameters are involved in PreGC, T_{block} indicating the ratio of free blocks and T_{page} indicating the ratio of valid pages. The workflow of PreGC performs as follows. First, PreGC judges whether the current number of free blocks is less than T_{block}. When this condition is satisfied, the victim block with the least valid pages would be determined according to the greedy algorithm. Then, the valid page ratio T_{page} in this block is further detected. Once the valid page ratio is less than this threshold, the current system status would be judged. Once system becomes idle, one valid page in the victim block would be migrated. When the first migration is finished, system status should be judged again to avoid delaying subsequent requests for long. Moreover, the valid page ratio would also be re-checked again. Thus, the conditions to stop PreGC can be triggered when the system becomes busy or when the valid page ratio is larger than T_{page}.

From the above workflow, we can find that the effectiveness of PreGC largely depends on system idle time as well as the pre-migration numbers. Thus, it would be evaluated comprehensively with multiple workloads having varied system idle time and with multiple parameter settings of T_{block} and T_{page} as the sensitivity study. Details of the evaluation would be presented in Section 5.

4.3. Cooperating with Normal GC

PreGC is a novel method to improve the performance of SSD by working together with GC and is actually not a replacement for existing GC methods that we call normal GC in this paper. Thus, PreGC is orthogonal with normal GC methods. This section presents how PreGC assists the normal GC to reduce page migrations. PreGC is often used before the normal GC on the victim block, as shown in Figure 7. In the period of 3D SSDs in Figure 7, PreGC and normal GC are both used. When the system is idle, part of the pages in a victim block are migrated during the yellow time slot. Then, the system becomes busy; as shown in the dark gray time slot, the migrations are stopped because of the system status. When the system becomes idle again, pre-migrations begin again. In this invoking, PreGC is stopped because that valid page ratio is satisfied. Consequently, normal GC is invoked and normal page migrations occur. From the changes of valid page distribution among several blocks, as shown in Figure 7, PreGC actually increases the number of valid pages. This also means that PreGC increases the extra write number for the case that valid pages are updated during the period between PreGC and normal GC. Thus, PreGC would induce write amplification, which also would be evaluated in Section 5.

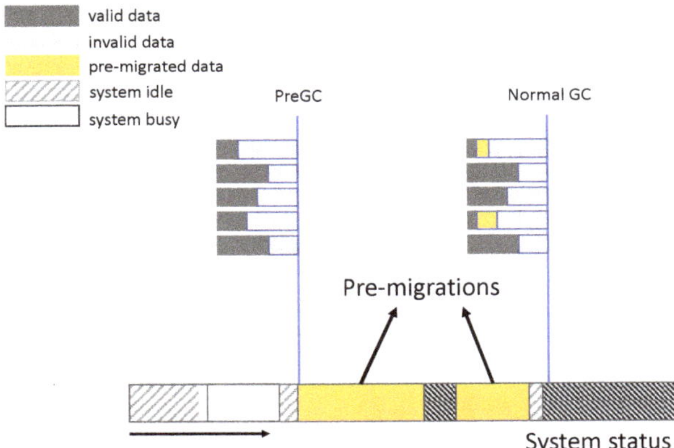

Figure 7. The cooperation between PreGC and normal GC. The box on the lower side of the figure represents the system status progress bar in the SSD, while the box on the upper side represents the page status. The figure shows the system status that will trigger PreGC and Normal GC as well as the current page status and the PreGC process that occurs between them.

5. Experiment and Evaluation

This section first describes the experiment platform and parameter configurations to evaluate our proposed PreGC. Then, the experimental results about performance and overhead of PreGC are shown and analyzed under five real-world workloads by comparing with the original GC method.

5.1. Experiment Setup

The experiment designed for PreGC evaluation is illustrated from the following four aspects. First, SSD configurations using the SSDsim simulator [37] are presented and the five real-world workloads are introduced. Then, the parameters settings in our experiment and sensitivity study are described. Lastly, we compare methods to evaluate the proposed PreGC method.

SSD configurations: The proposed PreGC method was integrated into the controller of 3D SSDs, and all experiments were conducted on a flash simulator named SSDsim [37], which is a reliable platform that has been widely used in many research works about SSDs [14,38,39].

Real-world workloads: To evaluate the effectiveness of PreGC on performance cliff and tail latency reduction, five real-world workloads with different features were chosen from Umass [40], as listed in Table 2. In our experiment, the duration of these workloads was about 18 hours.

Parameter settings: There were two thresholds involved in the PreGC flow chart, as illustrated earlier, which are the free block ratio threshold Tblock used to invoke page migrations in PreGC and the valid page ratio threshold T_{page} used to determine whether to proceed PreGC. By conducting a series of threshold value tests, we determined Tblock to be 11% and T_{page} to be 10% for all workloads. The trigger condition of normal GC is when the free block ratio reaches to 10%.

Compared methods: Our PreGC method is designed to assist the traditional GC methods, and we are the first to propose such a GC assistance from the aspect of page migrations. Thus, we compare the performance and overhead of SSD systems with and without ProGC together with the original GC method, and the excellent partial GC method GFTL. Moreover, we combined PreGC and GFTL to prove that our approach can work with other methods. The four compared methods are denoted as PreGC, Original, GFTL, and GFTL after PreGC.

It is worth mentioning that the comparison of the methods from GFTL and PreGC shows in Figure 8. The GFTL method divides the GC into several operations with a required time less than or equal to one erase latency after the GC condition is triggered and executes it one by one in the request interval, which is equivalent to delaying the normal foreground GC into a background GC to hide its latency, so it also requires a large amount of space as a buffer, for example, 16% in this experiment. The PreGC we proposed was to migrate valid pages of to be erased blocks ahead of time before the GC condition was triggered and to move one page at a time, thus reducing the current GC latency and avoiding blocking I/O for too long. PreGC does not interfere with normal GC operation because the GC operation is indispensable although it has some bad effects. In summary, PreGC has the following advantages: First, it does not interfere with the execution of normal GC but cooperates with it. Second, no additional buffer space is required. Finally, the time granularity of the step-by-step operation is smaller and more flexible.

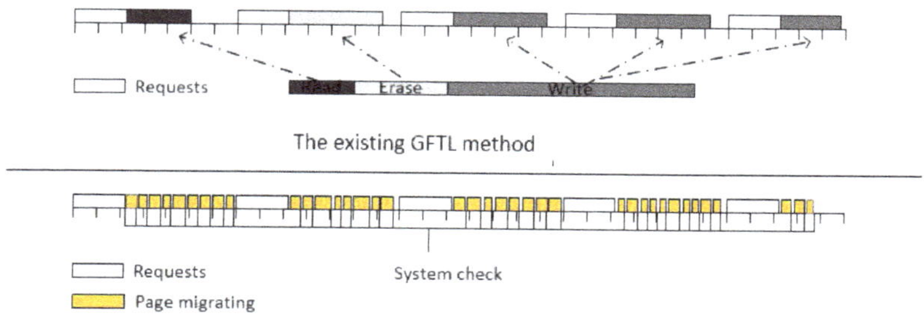

Figure 8. Comparison of two methods. The box in the figure represents the non-idle system state, and different colors indicate different states. The upper side of the figure shows the existing GFTL method, while the lower side shows the PreGC method proposed in this paper.

5.2. Results and Analysis

We first analyze the results of PreGC on normal page migration, which indicates the number of migrated pages when GC happens. As PreGC migrates some valid pages in advance, page migrations when GC happens are reduced, noting that our PreGC method does not reduce the overall migrated pages. We call page migrations in GC normal. Details about the reduction are presented in Table 4. Then, the performance results including the prorformance cliff phenomenon and tail latency after pre-migrating valid pages are presented to verify the effectiveness of PreGC. Moreover, the overhead of PreGC on the write amplification is also evaluated. Lastly, the workload characteristics are discussed in which PreGC can play the role more effectively.

Table 4. Page migration statistics.

Trace vs. Stat	Mig w/o PreGC	Mig w/ PreGC	Reduction	N_{PreGC}	$PreMIG$
usr0	44	31	29.5%	13,095	63
src0	60	42	30%	10,351	29
hm0	22	13	40.9%	5763	27
ts0	17	10	41.2%	11,851	23
rsrch0	25	14	44%	9116	25
wdev0	16	10	37.5%	1926	44
Avg.	30.6	20	34.6%	8684	35.2

5.2.1. The Number of Normally Migrated Pages in GC

In order to show the effect PreGC on page migrations, the average number of normally migrated pages is computed as Equation (1), in which MIG_{GC} is the totally migrated pages when GC happens and the N_{GC} represents the overall GC number. Moreover, the average number of pre-migrated pages for each workload computed according to Equation (2), in which MIG_{PreGC} represents the total page migrations induced by PreGC and N_{PreGC} indicates the overall number of PreGC invoking.

$$MIG_{average} = \frac{MIG_{GC}}{N_{GC}} \tag{1}$$

$$PreMIG_{average} = \frac{MIG_{PreGC}}{N_{PreGC}} \tag{2}$$

The comparison results without and with PreGC, the numbers of invoked PreGC, and the average pre-migration numbers by PreGC are presented in Table 4. According to these results, we can first find that the number of migrated pages are different for workloads. This is because that the situations that invoke PreGC for each workload are different from each other. It depends on the number of overall GC during the investigated period of this workload and mainly depends on the access density of workloads. The page reduction for workload rsrch0 is the highest, and the average migration reduction is 34.6% for these six workloads.

By analyzing the results of PreGC numbers and average pre-migrated page numbers, it can be found that pre-migrated page numbers are larger than normal page migration reduction and varies among workloads. These results are largely affected by the system idle time in workloads; due to that, page pre-migration can only be performed during the system is idle, the system status should be detected after each page pre-migration operation, and the next page pre-migration operation continues when the detection result of system status is idle. From Table 5, the average request interval time for workloads are varied, and it is one of the reasons for different pre-migrated page numbers between the workloads.

Table 5. Statistics of six real-world workloads.

Trace	Method	MIG_{GC}	N_{GC}	$MIG_{average}$
usr0	Original	477,346	15,575	30.65
	PreGC	89,023	14,839	6.00
src0	Original	12,595	5498	2.29
	PreGC	6667	5467	1.22
ts0	Original	15,532	9212	1.69
	PreGC	11,074	9188	1.21
rsrch0	Original	8367	7580	1.10
	PreGC	7587	7576	1.00
fiu_web	Original	8,046,570	118,757	67.76
	PreGC	942,405	114,740	8.21
mds1	Original	2638	579	4.56
	PreGC	635	541	1.17
Average	Original	1,427,174.67	26,200.17	18.01
	PreGC	176,231.83	25,391.83	3.14

5.2.2. Performance Improvement

This section presents the performance results of the original and PreGC in terms of performance cliff and tail latency.

Performance cliff: In order to intuitively compare the performance results before and after applying our proposed PreGC method, the performance cliff for workload hm0 is shown in Figure 9, which corresponds to the investigated period in Figure 4. It can be seen that performance cliff is relieved by PreGC when compared with the original and GFTL. Detailed results would be presented in the following sections.

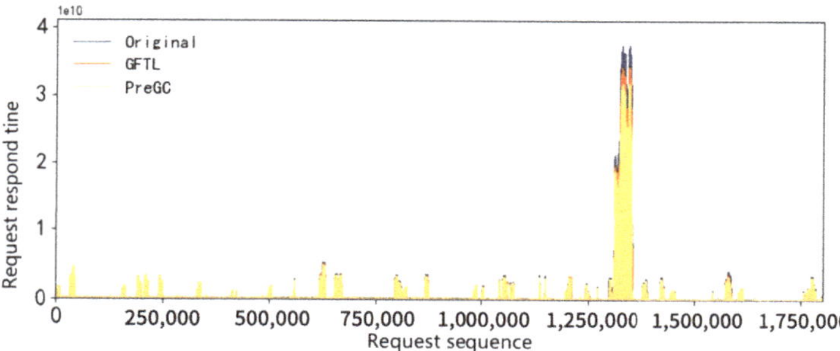

Figure 9. Comparison of process time. The figure shows the request response of the workload hm0, the abscissa is the request serial number, and the ordinate is the response time of the request.

Tail latency: Another quantitative evaluation of tail latency results with the 95th percentile and 99th percentile are presented in Figure 10. It can be observed that the two metrics have been significantly reduced by PreGC. The improvements in the 99th percentile are especially more obvious, which means that PreGC can bring about a more efficient reduction on the end of the long tail latency. Moreover, it can also found that the improvements are different among workloads. For the workload ts0, the latency is reduced most. On average, the tail latency can be reduced by 38.2%. These performance results show that our proposed PreGC can improve the SSD system performance and can relieve the performance cliff problem as well as long tail latency is induced by GC.

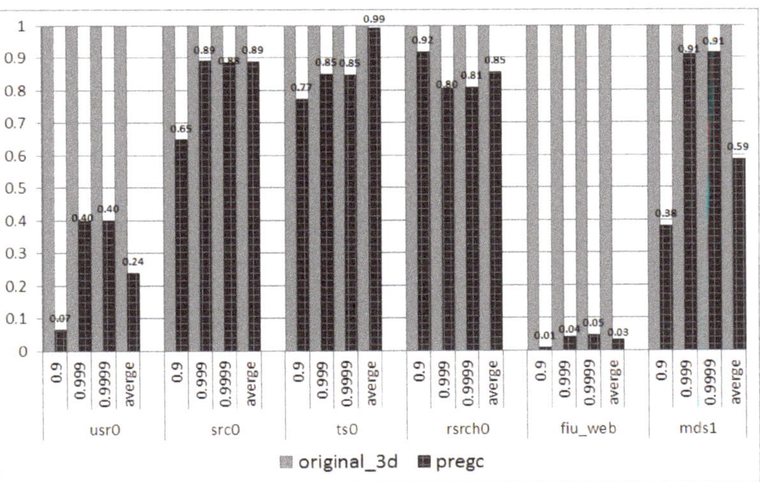

Figure 10. Comparison of tail latency related to GC. The figure shows the the normalized comparison result of the tail latancy of requests that may be affected by GC in original 3D SSDs and 3D SSDs with PreGC. Among them, based on the results of original 3D SSDs, the request tail latency of 2D SSDs is 50% less than that of 3D SSDs with PreGC on average.

5.2.3. Overhead on Write Amplification

As PreGC would migrate valid pages in advance before normal GC is invoked, the migrated pages might be updated during the pre-migration period and the victim block chosen in PreGC may not be the victim block in normal GC. Thus, PreGC would induce an extra write amplification, the results of which are shown in Figure 11. From the results, we can see that the write amplification for several traces is high but others are not. This is also decided by the characteristics of workloads. However, the average write amplification is under 1%, which can be negligible.

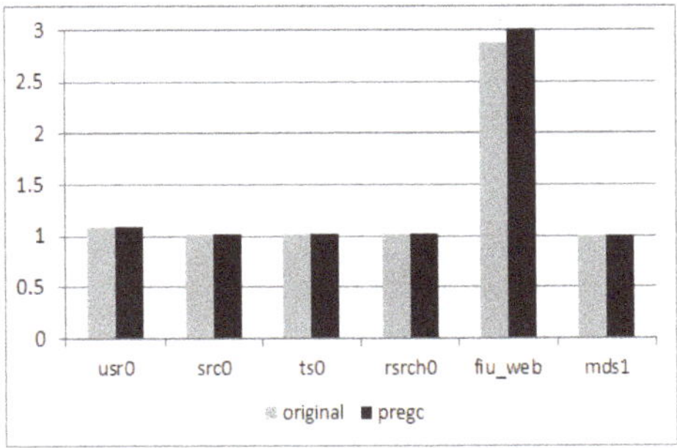

Figure 11. Write amplication contrast. The figure shows the comparison of the write amplification factor of original 3D SSDs and 3D SSDs with PreGC for different workloads.

5.2.4. Sensitivity Study

The above results have already verified the effectiveness of our proposed PreGC method under specific parameters. This section presents the performance result for more settings on key parameters in our implementation. Figures 12 and 13 show the comprehensive results when setting the threshold on free block proportion (T_{block}), and valid page ratio in a block (T_{page}). According to the results, three conclusions can be made. First, when T_{block} increases below a certain value, the tail latency decreases. However, when T_{block} exceeds a value, such as 10.75% that can be seen in the figure, the tail latency increases as the T_{block} increases. This is because, initially, an increase in T_{block} means that the PreGC threshold is easier to reach and it is easier to trigger PreGC to migrate the valid page in advance, thereby reducing GC latency and further reducing tail latency.

However, if the value continues to increase after a suitable value, it will cause the valid pages to be migrated too early, which will lead to a lot of invalid data to be generated and results in more GC; then, the request may be suspended for a longer period of time, which makes the tail latency longer. Second, the 99th tail latency increases as the value of T_{page} increases, but the 95th tail delay reaches a local peak when T_{page} is 10. This is because an increase in T_{page} means that the number of pages that a PreGC needs to pre-migrate increases, so that a more severe write amplification works in conjunction with a smaller number of valid pages included in the victim block in the short term, causing the above-described change in tail latency. These parameters can be adjusted in practice according to the performance requirement.

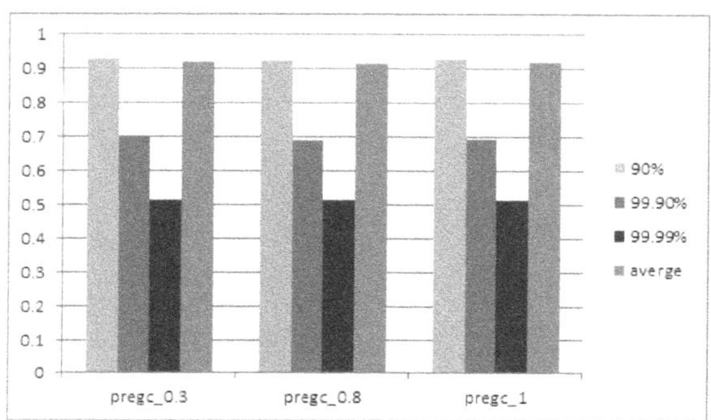

Figure 12. Sensitivity study results on the parameter of free block proportion to invoke PreGC.

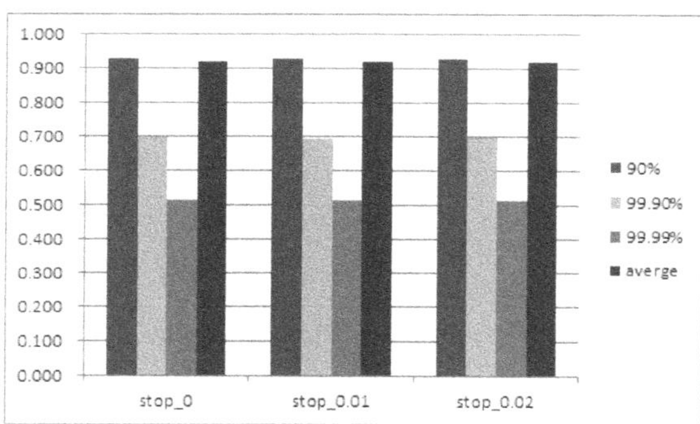

Figure 13. Sensitivity study results on the parameter of valid page ratio to invoke PreGC.

5.3. Discussion

Our PreGC method provides an assistance to existing GC methods and are orthogonal with many GC optimization methods. The pre-migrations would happen between the PreGC invoking time and normal GC invoking time when SSD system is idle. Thus, the effectiveness of PreGC can be largely exploited for workloads that have long system idle time close to the GC invoking time. Although PreGC can relieve performance improvements on tail latency, the problem of write amplification caused by the pre-migration of valid pages, that is, the amount of data actually written in the SSDs, is many times the amount of data that the host requests to write. Although it is inevitable for pre-migrations to cause write amplification, PreGC applies a mechanism to stop it in time to alleviate the problem. Therefore, the write amplification brought about by this method is within the small range. The other overhead is to store two thresholds for triggering and stopping PreGC. As the two parameters only take up a small space, the storage overhead caused by our method can be ignored.

6. Conclusions

In order to satisfy the increased concerns about SSD performance, this paper studied GC performance, which closely relates to system performance, in the view of performance cliff and tail latency. Several observations have been found from our preliminary experiments. The root cause of performance cliff, increased page migrations, has been figured out. A new garbage collection method, PreGC, is proposed to invoke partial page migrations in advance, which can reduce the GC latency effectively. Experimental results have shown the effectiveness of PreGC. As our method is also suitable for optimizing wear leveling schemes, we will study this problem in our future work.

Author Contributions: This contribution of this paper from authors is as follows: Y.D. is responsible for conceptualization, methodology, investigation, writing on original draft preparation, supervision, project administration and funding acquisition; W.L. is responsible for data curation, software, validation and formal analysis; R.A. is responsible for resources and writing—review and editing; Y.G. is responsible for visualization. All authors have read and agreed to the published version of the manuscript.

Funding: This research was funded by National Natural Science Foundation of China grant number 61802287.

Conflicts of Interest: The authors declare no conflicts of interest.

References

1. ATC. 3D NAND Stacking Memory Cells. 2018. Available online: https://www.atpinc.com/blog/3d-nand-ssd-sd-flash-memory-storage-what-is (acceseed on 10 November 2020).
2. Kang, D.; Jeong, W.; Kim, C.; Kim, D.; Cho, Y.S.; Kang, K.; Ryu, J.; Kang, K.; Lee, S.; Kim, W.; et al. 256 Gb 3 b/Cell V-nand Flash Memory With 48 Stacked WL Layers. *IEEE J. Solid-State Circuit.* **2017**, *52*, 210–217. [CrossRef]
3. Park, K.; Nam, S.; Kim, D.; Kwak, P.; Lee, D.; Choi, Y.; Choi, M.; Kwak, D.; Kim, D.; Kim, M.; et al. Three-Dimensional 128 Gb MLC Vertical nand Flash Memory With 24-WL Stacked Layers and 50 MB/s High-Speed Programming. *IEEE J. Solid-State Circuit.* **2015**, *50*, 204–213. [CrossRef]
4. Deguchi, Y.; Takeuchi, K. 3D-NAND Flash Solid-State Drive (SSD) for Deep Neural Network Weight Storage of IoT Edge Devices with 700x Data-Retention Lifetime Extention. In Proceedings of the 2018 IEEE International Memory Workshop (IMW), Kyoto, Japan, 13–16 May 2018; pp. 1–4. [CrossRef]
5. Kang, Y.; Zhang, X.; Shao, Z.; Chen, R.; Wang, Y. A reliability enhanced video storage architecture in hybrid SLC/MLC NAND flash memory. *J. Syst. Archit.* **2018**, *88*, 33–42. [CrossRef]
6. Li, Q.; Shi, L.; Yang, J.; Zhang, Y.; Xue, C.J. Leveraging Approximate Data for Robust Flash Storage. In Proceedings of the 56th Annual Design Automation Conference, ACM, Las Vegas, NV, USA, 2 June 2019; p. 215.
7. Ji, C.; Pan, R.; Chang, L.P.; Shi, L.; Zhu, Z.; Liang, Y.; Kuo, T.W.; Xue, C.J. Inspection and Characterization of App File Usage in Mobile Devices. *ACM Trans. Storage* **2020**, *16*. [CrossRef]
8. Ji, C.; Chang, L.P.; Pan, R.; Wu, C.; Gao, C.; Shi, L.; Kuo, T.W.; Xue, C.J. Pattern-Guided File Compression with User-Experience Enhancement for Log-Structured File System on Mobile Devices. In Proceedings of the 19th USENIX Conference on File and Storage Technologies (FAST 21), USENIX Association, Virtual Event, 23–25 February 2021; pp. 127–140.
9. Misra, P.A.; Borge, M.F.; Goiri, I.N.; Lebeck, A.R.; Zwaenepoel, W.; Bianchini, R. Managing Tail Latency in Datacenter-Scale File Systems Under Production Constraints. In Proceedings of the Fourteenth EuroSys Conference 2019, EuroSys'19, Dresden, Germany, 25–28 March 2019; ACM: New York, NY, USA, 2019; pp. 17:1–17:15. [CrossRef]
10. Paik, J.; Cho, E.; Jin, R.; Chung, T. Selective-Delay Garbage Collection Mechanism for Read Operations in Multichannel Flash-Based Storage Devices. *IEEE Trans. Consum. Electr.* **2018**, *64*, 118–126. [CrossRef]
11. Wu, S.; Zhu, W.; Liu, G.; Jiang, H.; Mao, B. GC-Aware Request Steering with Improved Performance and Reliability for SSD-Based RAIDs. In Proceedings of the 2018 IEEE International Parallel and Distributed Processing Symposium (IPDPS), Vancouver, BC, Canada, 21–25 May 2018; pp. 296–305. [CrossRef]
12. Chen, S.; Chang, Y.; Liang, Y.; Wei, H.; Shih, W. An Erase Efficiency Boosting Strategy for 3D Charge Trap NAND Flash. *IEEE Trans. Comput.* **2018**, *67*, 1246–1258. [CrossRef]
13. Cui, J.; Zhang, Y.; Huang, J.; Wu, W.; Yang, J. ShadowGC: Cooperative garbage collection with multi-level buffer for performance improvement in NAND flash-based SSDs. In Proceedings of the 2018 Design, Automation Test in Europe Conference Exhibition (DATE), Dresden, Germany, 19–23 March 2018; pp. 1247–1252. [CrossRef]

14. Yan, S.; Li, H.; Hao, M.; Tong, M.H.; Sundararaman, S.; Chien, A.A.; Gunawi, H.S. Tiny-Tail Flash: Near-Perfect Elimination of Garbage Collection Tail Latencies in NAND SSDs. *ACM Trans. Storage* **2017**, *13*, 22:1–22:26. [CrossRef]
15. Choi, W.; Jung, M.; Kandemir, M.; Das, C. Parallelizing Garbage Collection with I/O to Improve Flash Resource Utilization. In Proceedings of the 27th International Symposium on High-Performance Parallel and Distributed Computing, HPDC'18, Tempe, AZ, USA, 11–15 June 2018; ; ACM: New York, NY, USA, 2018; pp. 243–254. [CrossRef]
16. Guo, J.; Hu, Y.; Mao, B.; Wu, S. Parallelism and Garbage Collection Aware I/O Scheduler with Improved SSD Performance. In Proceedings of the 2017 IEEE International Parallel and Distributed Processing Symposium (IPDPS), Orlando, FL, USA, 29 May–2 June 2017; pp. 1184–1193. [CrossRef]
17. Shahidi, N.; Kandemir, M.T. CachedGC: Cache-Assisted Garbage Collection in Modern Solid State Drives. In Proceedings of the 2018 IEEE 26th International Symposium on Modeling, Analysis, and Simulation of Computer and Telecommunication Systems (MASCOTS), Milwaukee, WI, USA, 25–28 September 2018; pp. 79–86. [CrossRef]
18. Luo, Y.; Ghose, S.; Cai, Y.; Haratsch, E.F.; Mutlu, O. Improving 3D NAND flash memory lifetime by tolerating early retention loss and process variation. In*Abstracts of the 2018 ACM International Conference on Measurement and Modeling of Computer Systems*; ACM: New York, NY, USA, 2018; pp. 106–106.
19. Yang, M.; Chang, Y.; Tsao, C.; Huang, P.; Chang, Y.; Kuo, T. Garbage collection and wear leveling for flash memory: Past and future. In Proceedings of the 2014 International Conference on Smart Computing, Hong Kong, China, 3–5 November 2014; pp. 66–73. [CrossRef]
20. Goda, A.; Parat, K. Scaling directions for 2D and 3D NAND cells. In Proceedings of the 2012 International Electron Devices Meeting, San Francisco, CA, USA, 10–13 December 2012; pp. 2.1.1–2.1.4. [CrossRef]
21. Kim, Y.; Mateescu, R.; Song, S.; Bandic, Z.; Kumar, B.V.K.V. Coding scheme for 3D vertical flash memory. In Proceedings of the 2015 IEEE International Conference on Communications (ICC), London, UK, 8–12 June 2015; pp. 264–270. [CrossRef]
22. Wu, F.; Lu, Z.; Zhou, Y.; He, X.; Tan, Z.; Xie, C. OSPADA: One-Shot Programming Aware Data Allocation Policy to Improve 3D NAND Flash Read Performance. In Proceedings of the 2018 IEEE 36th International Conference on Computer Design (ICCD), Orlando, FL, USA, 7–10 October 2018; pp. 51–58. [CrossRef]
23. Shihab, M.M.; Zhang, J.; Jung, M.; Kandemir, M. ReveNAND: A Fast-Drift-Aware Resilient 3D NAND Flash Design. *ACM Trans. Archit. Code Optim.* **2018**, *15*, 17:1–17:26. [CrossRef]
24. Cui, J.; Zhang, Y.; Shi, L.; Xue, C.J.; Wu, W.; Yang, J. ApproxFTL: On the Performance and Lifetime Improvement of 3-D NAND Flash-Based SSDs. *IEEE Trans. Comput. Aided Design Integr. Circuit. Syst.* **2018**, *37*, 1957–1970. [CrossRef]
25. Pletka, R.; Ioannou, N.; Papandreou, N.; Parnell, T.; Tomic, S. Enabling 3D-TLC NAND Flash in Enterprise Storage Systems. *ERCIM NEWS* **2018**, 48–49. Available online: https://ercim-news.ercim.eu/en113/r-i/enabling-3d-tlc-nand-flash-in-enterprise-storage-syste (accessed on 1 June 2021).
26. Ho, C.; Li, Y.; Chang, Y.; Chang, Y. Achieving Defect-Free Multilevel 3D Flash Memories with One-Shot Program Design. In Proceedings of the 2018 55th ACM/ESDA/IEEE Design Automation Conference (DAC), San Francisco, CA, USA, 24–28 June 2018; pp. 1–6. [CrossRef]
27. Zhang, M.; Wu, F.; Chen, X.; Du, Y.; Liu, W.; Zhao, Y.; Wan, J.; Xie, C. RBER Aware Multi-Sensing for Improving Read Performance of 3D MLC NAND Flash Memory. *IEEE Access* **2018**, *6*, 61934–61947. [CrossRef]
28. Chen, S.; Chen, Y.; Chang, Y.; Wei, H.; Shih, W. A Progressive Performance Boosting Strategy for 3-D Charge-Trap NAND Flash. *IEEE Trans. Very Large Scale Integr. Syst.* **2018**, *26*, 2322–2334. [CrossRef]
29. Xiong, Q.; Wu, F.; Lu, Z.; Zhu, Y.; Zhou, Y.; Chu, Y.; Xie, C.; Huang, P. Characterizing 3D Floating Gate NAND Flash: Observations, Analyses, and Implications. *ACM Trans. Storage* **2018**, *14*, 16:1–16:31. [CrossRef]
30. Wu, F.; Zhu, Y.; Xiong, Q.; Lu, Z.; Zhou, Y.; Kong, W.; Xie, C. Characterizing 3D Charge Trap NAND Flash: Observations, Analyses and Applications. In Proceedings of the 2018 IEEE 36th International Conference on Computer Design (ICCD), Orlando, FL, USA, 7–10 October 2018; pp. 381–388. [CrossRef]
31. Hung, C.; Chang, M.; Yang, Y.; Kuo, Y.; Lai, T.; Shen, S.; Hsu, J.; Hung, S.; Lue, H.; Shih, Y.; et al. Layer-Aware Program-and-Read Schemes for 3D Stackable Vertical-Gate BE-SONOS NAND Flash Against Cross-Layer Process Variations. *IEEE J. Solid State Circut.* **2015**, *50*, 1491–1501. [CrossRef]
32. Liu, C.Y.; Kotra, J.B.; Jung, M.; Kandemir, M.T.; Das, C.R. SOML Read: Rethinking the Read Operation Granularity of 3D NAND SSDs. In Proceedings of the Twenty-Fourth International Conference on Architectural Support for Programming Languages and Operating Systems, ASPLOS'19, Providence, RI, USA, 13–17 April 2019; ACM: New York, NY, USA, 2019; pp. 955–969. [CrossRef]
33. Wang, Y.; Dong, L.; Mao, R. P-Alloc: Process-Variation Tolerant Reliability Management for 3D Charge-Trapping Flash Memory. *ACM Trans. Embed. Comput. Syst.* **2017**, *16*, 142:1–142:19. [CrossRef]
34. Choudhuri, S.; Givargis, T. Deterministic Service Guarantees for NAND Flash using Partial Block Cleaning. In Proceedings of the 6th IEEE/ACM/IFIP International Conference on Hardware/Software Codesign and System Synthesis, Atlanta, GA, USA, 19–24 October 2009; Volume 4, pp. 19–24.
35. Jung, M.; Prabhakar, R.; Kandemir, M.T. Taking Garbage Collection Overheads Off the Critical Path in SSDs. In *ACM/IFIP/USENIX International Conference on Distributed Systems Platforms and Open Distributed Processing*; Springer: Berlin/Heidelberg, Germany, 2012.
36. Narayanan, D.; Donnelly, A.; Rowstron, A. Write Off-loading: Practical Power Management for Enterprise Storage. *Trans. Storage* **2008**, *4*, 10:1–10:23. [CrossRef]

37. Hu, Y.; Jiang, H.; Feng, D.; Tian, L.; Luo, H.; Zhang, S. Performance impact and interplay of SSD parallelism through advanced commands, allocation strategy and data granularity. In Proceedings of the International Conference on Supercomputing, Tucson, AZ, USA, 31 May–4 June 2011; pp. 96–107.
38. Elyasi, N.; Arjomand, M.; Sivasubramaniam, A.; Kandemir, M.T.; Das, C.R.; Jung, M. Exploiting intra-request slack to improve SSD performance. *ACM SIGARCH Comput. Arch. News* **2017**, *45*, 375–388. [CrossRef]
39. Zhang, W.; Cao, Q.; Jiang, H.; Yao, J. PA-SSD: A Page-Type Aware TLC SSD for Improved Write/Read Performance and Storage Efficiency. In Proceedings of the 2018 International Conference on Supercomputing, Beijing, China, 12–15 June 2018; pp. 22–32.
40. UMass. UMass Trace Repository. 2007. Available online: http://traces.cs.umass.edu/index.php/Storage/Storage (accessed on 10 June 2021).

 micromachines

MDPI

Article

Artificial Neural Network Assisted Error Correction for MLC NAND Flash Memory

Ruiquan He [1], Haihua Hu [2], Chunru Xiong [1] and Guojun Han [1,*]

[1] ZTE School of Information Technology, Xinyu University, Xinyu 338025, China; Ray_HRQ@outlook.com (R.H.); xcrmcu@163.com (C.X.)
[2] School of Information Engineering, Guangdong University of Technology, Guangzhou 510006, China; haihua@mail2.gdut.edu.cn
* Correspondence: gjhan@gdut.edu.cn

Abstract: The multilevel per cell technology and continued scaling down process technology significantly improves the storage density of NAND flash memory but also brings about a challenge in that data reliability degrades due to the serious noise. To ensure the data reliability, many noise mitigation technologies have been proposed. However, they only mitigate one of the noises of the NAND flash memory channel. In this paper, we consider all the main noises and present a novel neural network-assisted error correction (ANNAEC) scheme to increase the reliability of multi-level cell (MLC) NAND flash memory. To avoid using retention time as an input parameter of the neural network, we propose a relative log-likelihood ratio (LLR) to estimate the actual LLR. Then, we transform the bit detection into a clustering problem and propose to employ a neural network to learn the error characteristics of the NAND flash memory channel. Therefore, the trained neural network has optimized performances of bit error detection. Simulation results show that our proposed scheme can significantly improve the performance of the bit error detection and increase the endurance of NAND flash memory.

Keywords: NAND flash memory; artificial neural network; error correction code; reliability

Citation: He, R.; Hu, H.; Xiong, C.; Han, G. Artificial Neural Network Assisted Error Correction for MLC NAND Flash Memory. *Micromachines* **2021**, *12*, 879. https://doi.org/10.3390/mi12080879

Academic Editors: Cristian Zambelli and Rino Micheloni

Received: 30 June 2021
Accepted: 20 July 2021
Published: 27 July 2021

1. Introduction

NAND flash memories have been widely used in smartphones, personal computers, data centers, etc. Thanks to these two key technologies: (1) continued scaling down process technology and (2) multilevel (e.g., MLC, TLC) cell data coding, the storage density of a NAND flash memory has been significantly increased over previous decades [1]. However, these two key technologies bring about a challenge in that the data stored in NAND flash memory may suffer from low reliability [2–4]. Furthermore, there are two major sources of noise in flash memory: cell-to-cell interference (CCI) and retention noise. Numerous works have been proposed to mitigate noises in NAND flash memory. For example, the data post compensation and predistortion technique [5] and detector design using a neighbor-a-priori information technique [6] exploit the a-priori information of the neighboring cells to mitigate the CCI. However, when considering retention noise, the voltage offset of flash memory cell tends to become unknown. It may be hard to use the a-priori information of the neighboring cells to compensate for the voltage shift caused by CCI. In addition, the CCI removal technique proposed by Lin [7] suffers from a similar problem in that the proposed technique ignores the impact of noise. In addition, Reference [8] proposed a retention-aware belief-propagation (BP) decoding scheme to mitigate the retention noise effect but did not take CCI into consideration.

Against the above background, the recent advances in neural networks and machine learning provide a new perspective to increase the reliability of MLC NAND flash memory. The key idea of the neural network is to learn an optimal network model from the massive training data, instead of using a definitive algorithm that is derived from a pre-defined model [9]. A pioneering work is reported in [10,11], which utilizes an artificial neural network to predict the threshold voltage distribution of NAND flash memory. In the pretesting, the above method assumes that the prior information of the retention time is informed in advance. When the flash controller is powered off, we cannot obtain the retention time.

In this paper, we use the neural network to learn an optimal network model to detect the bits errors in the cells that are disturbed by both CCI and retention noise and propose a neural network-assisted error correction scheme. However, it is difficult to record the retention time in a practical system, which means that accurate LLR values cannot be calculated. Therefore, we propose using relative LLR to estimate the actual LLR. The relative LLR is affected little by retention time, so we do not require retention time as an input parameter of the neural network.

In this paper, we first model the threshold voltage distribution as a Gaussian mixture model, which is fairly close to the voltage distribution of the practical NAND flash memory, and we calculate the LLR of the theoretical threshold distribution using a quantization scheme. Then, the corresponding LLR of the actual threshold distribution is mapped according to the relative position of the optimal reading reference voltage. It is found that this idea makes the relative LLR values remain relatively steady throughout retention time, which allows us to avoid using retention time as an input parameter of the neural network. Finally, using the relative LLR to estimate the actual LLR, we train the neural network and use the trained network to recovery the bits that may be wrongly detected in the soft-decision detection or hard-decision detection.

The rest of this paper is organized as follows. The flash channel model is presented in Section 2. Section 3 introduces our proposed ANNAEC scheme. Numerical simulation results are presented in Section 4. The conclusions are drawn in Section 5.

2. Channel Model

Without loss of generality, the proposed ANNAEC is performed over a model-based MLC NAND flash memory. Based on [5,8,12], we can model threshold voltage, V_{th}, by

$$V_{th} = V + n_{RTN} + \triangle V_{CCI} - n_{retention}, \tag{1}$$

where V denotes the desired voltage level, n_{RTN} denotes random telegraph noise (RTN), $\triangle V_{CCI}$ denotes the shift caused by *CCI* noise, and $n_{retention}$ denotes retention noise.

2.1. The Voltage Distribution of Programmed and Erased Cell

The number of charges in the NAND flash memory cell can be altered in the program and erase operation. It is well known that before being programmed, a flash memory cell must be erased. In the erase operation, the charges in the memory cell are removed from the floating gate, and the threshold voltage of the erased cell will be set to the lowest voltage. The threshold voltage distribution of an erased cell follows a Gaussian distribution, which is given by

$$p_e(x) = \frac{1}{\sigma_e\sqrt{2\pi}} e^{-\frac{(x-\mu_e)^2}{2\sigma_e^2}} = \mathcal{N}(\mu_e, \sigma_e^2), \tag{2}$$

where σ_e and μ_e are the standard deviation and the mean of the threshold voltage of the erased cell, respectively.

According to [5,8], the threshold voltage of a programmed cell follows a Gaussian distribution shown below:

$$p_p(x) = \frac{1}{\sigma_p\sqrt{2\pi}} e^{-\frac{(x-\mu_p)^2}{2\sigma_p^2}} = \mathcal{N}(\mu_p, \sigma_p^2), \tag{3}$$

where σ_p and $\mu_p \in \{\mu_{P_{01}}, \mu_{P_{00}}, \mu_{P_{10}}\}$ are the standard deviation and the mean of the threshold voltage of a programmed cell.

2.2. RTN

The electron capture and emission at the floating gate near the interface generate RTN, which is greatly impacted by flash memory P/E cycles [13]. As P/E cycles increase, the tunnel oxide of the floating gate transistor is gradually damaged and generates charge trapping in the oxide and interface states. RTN leads to a random fluctuation of cell threshold voltage and widens the voltage distribution. Hence, RTN is modeled with a Gaussian-like distribution [8], given as

$$p_r(x) = \frac{1}{\sigma_r\sqrt{2\pi}}e^{-\frac{x^2}{2\sigma_r^2}} = \mathcal{N}(0,\ \sigma_r^2), \tag{4}$$

where $\sigma_r = 0.00027 \times PE^{0.62}$, denotes the noise standard deviation.

2.3. CCI

Because of the parasitic capacitance-coupling effect among adjacent cells in flash memory, the threshold voltage of the victim cell increases as the threshold voltage of an adjacent cell increases. The immediate adjacent cells are the major noise source of the CCI. We consider an all bit-line structure. As shown in Figure 1, when the (k+1)-th wordline (WL) has been programmed, the cell on the k-th WL can be programmed. Hence, the victim cell is influenced by three immediate adjacent cells. The threshold-voltage shift of the victim cell can be modeled as a linear combination of the threshold voltage changes of those immediate adjacent cells. We can estimate the threshold-voltage shift caused by CCI as

$$\triangle V_{victim} = \sum_n (\triangle V_t^{(n)} \cdot \gamma^{(n)}), \tag{5}$$

where $\triangle V_t^{(n)}$ is the change of an immediate adjacent cell, which is programmed after the victim cell and $\gamma^{(n)}$ represents the coupling ratio. We assume the vertical and the diagonal coupling ratio are γ_y and γ_{xy}, respectively. According to the cell-to-cell coupling strength factor s, we can set $\gamma_y = 0.08s$ and $\gamma_{xy} = 0.006s$ [12].

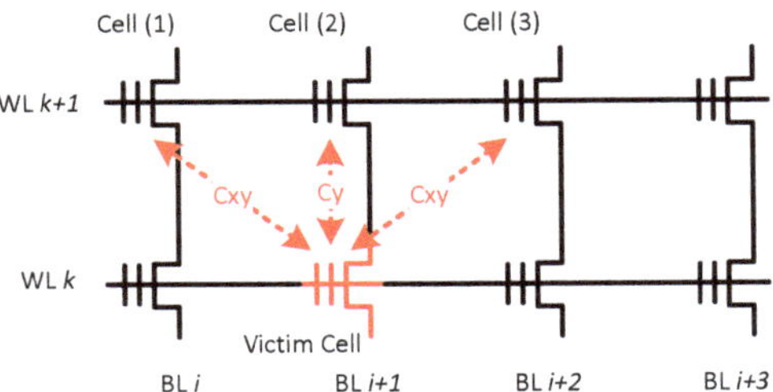

Figure 1. Illustration of the parasitic coupling capacitances among adjacent cells.

2.4. Retention

After a cell is programmed, the number of charges in the NAND flash memory cell continually reduce over time due to trap-assisted tunneling and charge detrapping [1]. Retention noise is modeled as a Gaussian distribution, i.e., $p_t(x) = \mathcal{N}(\mu_t, \sigma_t^2) = \frac{1}{\sqrt{2\pi}\sigma_t}e^{-\frac{(x-\mu_t)^2}{2\sigma_t^2}}$. The mean μ_t, and the standard deviation σ_t, are given by

$$\mu_t = \triangle V_t[A_t(PE)^{\alpha_i} + B_t(PE)^{\alpha_o}]\log(1+T), \tag{6}$$

$$\sigma_t = 0.3|\mu_t|, \tag{7}$$

where $\triangle V_t$ is the cell voltage change before and after being programmed, T donates memory retention time and PE is the number of PE cycles.

The conditional probability distribution function of the threshold voltage after being disturbed by RTN, CCI and retention are given as follows:

$$p(V_{th}|k \in \{11,01,00,01\}) = \frac{1}{64}[\mathcal{N}(\mu_k - \mu_t, \sigma_k^2 + \sigma_t^2 + \sigma_r^2) + A + B + C], \tag{8}$$

$$A = \sum_{\mu_p}[2\mathcal{N}(\gamma_{xy}(\mu_p - \mu_e) + \mu_k - \mu_t, \gamma_{xy}^2(\sigma_p^2 + \sigma_e^2 + 2\sigma_r^2) + \sigma_k^2 + \sigma_t^2) + \mathcal{N}(\gamma_y(\mu_p - \mu_e) + \mu_k - \mu_t, \gamma_y^2(\sigma_p^2 + \sigma_e^2 + 2\sigma_r^2) + \sigma_k^2 + \sigma_t^2)], \tag{9}$$

$$B = \sum_{\mu_p^{(1)}}\sum_{\mu_p^{(2)}}\sum_{\mu_p^{(3)}}\mathcal{N}(\gamma_{xy}(\mu_p^{(1)} + \mu_p^{(2)} - 2\mu_e) + \gamma_y(\mu_p^{(2)} - \mu_e) + \mu_k - \mu_t, (2\gamma_{xy}^2 + \gamma_y^2)(\sigma_p^2 + \sigma_e^2 + 2\sigma_r^2) + \sigma_k^2 + \sigma_t^2), \tag{10}$$

$$\begin{aligned} C = &\sum_{\mu_p^{(1)}}\sum_{\mu_p^{(2)}}\mathcal{N}(\gamma_{xy}(\mu_p^{(1)} - \mu_e) + \gamma_y(\mu_p^{(2)} - \mu_e) + \mu_k - \mu_t, (\gamma_{xy}^2 + \gamma_y^2)(\sigma_p^2 + \sigma_e^2 + 2\sigma_r^2 + \sigma_k^2 + \sigma_t^2)) \\ &+ \sum_{\mu_p^{(2)}}\sum_{\mu_p^{(3)}}\mathcal{N}(\gamma_{xy}(\mu_p^{(3)} - \mu_e) + \gamma_y(\mu_p^{(2)} - \mu_e) + \mu_k - \mu_t, (\gamma_{xy}^2 + \gamma_y^2)(\sigma_p^2 + \sigma_e^2 + 2\sigma_r^2) + \sigma_k^2 + \sigma_t^2) \\ &+ \sum_{\mu_p^{(1)}}\sum_{\mu_p^{(3)}}\mathcal{N}(\gamma_{xy}(\mu_p^{(1)} + \mu_p^{(2)} - 2\mu_e) + \mu_k - \mu_t, 2\gamma_{xy}^2(\sigma_p^2 + \sigma_e^2 + 2\sigma_r^2) + \sigma_k^2 + \sigma_t^2), \end{aligned} \tag{11}$$

where $\mu_p^{(1)}$, $\mu_p^{(2)}$ and $\mu_p^{(3)}$ are the means of cells 1–3, respectively, which are shown in Figure 2, μ_k and σ_k are the mean and standard deviation of the victim cell.

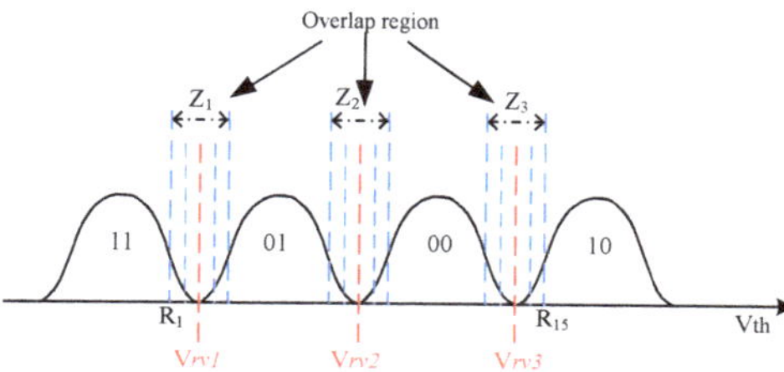

Figure 2. Illustration of 15-level uniform sensing quantization for multi-level cell (MLC) flash memory.

In this paper, we set the flash memory parameters as follows: $\mu_{p11} = 1.2$, $\mu_{p01} = 2.55$, $\mu_{p00} = 3$, $\mu_{p10} = 3.45$, $\sigma_p = 0.05$, $\sigma_e = 0.35$, $A_t = 0.000035$, $B_t = 0.000235$, $\alpha_i = 0.62$ and $\alpha_o = 0.30$.

3. Artificial Neural Network-Assisted Error Correction

In this section, we first present the idea of relative LLR calculation. Then we explain why an artificial neural network is useful for NAND flash memory. Finally, we introduce our proposed ANNAEC scheme.

3.1. Relative LLR

For soft decision belief-propagation (BP) decoding, a soft quantization scheme has been proposed. As an example, Figure 2 shows a 15-level uniform sensing quantization [12].

The overlap region is obtained by the entropy of the cell's threshold voltage [12,14]. When the threshold voltage falls into the range $(R_{n-1}, R_n]$, where R_n is the n-th reference voltage, $R_0 = -\infty$ and $R_{16} = +\infty$, the LLR values of the least significant bit (LSB) and the most significant bit (MSB) in the i-th cell can be calculated by (12) and (13), respectively:

$$LLR_{lsb}(R_{n-1}, R_n) = \log \frac{\int_{R_{n-1}}^{R_n} p(V_{th}|11) + p(V_{th}|01)\, \mathrm{d}x}{\int_{R_{n-1}}^{R_n} p(V_{th}|00) + p(V_{th}|10)\, \mathrm{d}x}, \tag{12}$$

$$LLR_{msb}(R_{n-1}, R_n) = \log \frac{\int_{R_{n-1}}^{R_n} p(V_{th}|11) + p(V_{th}|10)\, \mathrm{d}x}{\int_{R_{n-1}}^{R_n} p(V_{th}|01) + p(V_{th}|00)\, \mathrm{d}x}. \tag{13}$$

However, it may be hard to accurately calculate the LLR values due to the retention noise. Even though retention noise is modeled as Gaussian distribution, the mean and the standard deviation are random, since $\triangle V_t$ is random as described in (6) and (7). Furthermore, it is difficult to obtain accurate retention time in a practical system. To deal with those problems, we can estimate LLR, based on the relative reference voltage positions, given as

$$\begin{aligned} &LLR'_{lsb}(R_{n-1} - V_{rv} + V'_{rv}, R_n - V_{rv} + V'_{rv}) \\ &= \log \frac{\int_{R_{n-1}-V_{rv}+V'_{rv}}^{R_n-V_{rv}+V'_{rv}} p'(V_{th}|11) + p'(V_{th}|01)\, \mathrm{d}x}{\int_{R_{n-1}-V_{rv}+V'_{rv}}^{R_n-V_{rv}+V'_{rv}} p'(V_{th}|00) + p'(V_{th}|10)\, \mathrm{d}x}, \end{aligned} \tag{14}$$

$$\begin{aligned} &LLR'_{msb}(R_{n-1} - V_{rv} + V'_{rv}, R_n - V_{rv} + V'_{rv}) \\ &= \log \frac{\int_{R_{n-1}-V_{rv}+V'_{rv}}^{R_n-V_{rv}+V'_{rv}} p'(V_{th}|11) + p'(V_{th}|10)\, \mathrm{d}x}{\int_{R_{n-1}-V_{rv}+V'_{rv}}^{R_n-V_{rv}+V'_{rv}} p'(V_{th}|01) + p'(V_{th}|00)\, \mathrm{d}x}, \end{aligned} \tag{15}$$

where p' means that we estimate $\triangle V_t$ in Equations (6) and (7) as $\triangle V_t \approx \mu_k - \mu_e$, V_{rv} and V'_{rv} are the reference voltages of the actual threshold distribution and the theoretical threshold distribution, respectively, as shown in Figure 3, where V_{rv} is obtained by voltage optimization [1] and V'_{rv} is obtained by theoretical calculations, such as minimizing entropy of the cell's threshold voltage [12,14]. In (14) and (15), we first calculate the LLR of the theoretical threshold distribution using a quantization scheme. Then, the corresponding LLR of the actual threshold distribution is mapped according to the relative position of the optimal reference voltage.

We depict the relative LLR versus data retention time in Figure 4. The relative LLR values remain relatively steady, which allows the neural network to not require retention time as an input parameter. In addition, LLR calculation is offline in a flash memory controller [15]. It may be difficult for a controller to estimate the characteristics of the memory channel because online estimation leads to a significant increase in the power consumption and read latency of the flash controller. Therefore, the proposed relative LLR can estimate the actual LLR over a time range, which can also help reduce the number of LLR tables stored in the controller.

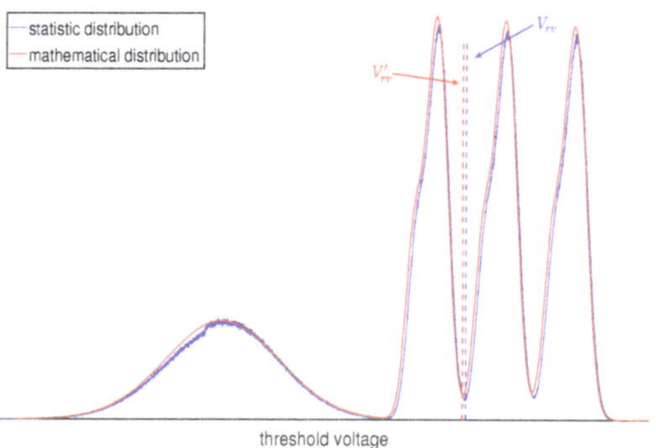

Figure 3. Illustration of the statistic distribution and mathematical distribution at $s = 1$ and $PE = 1K$.

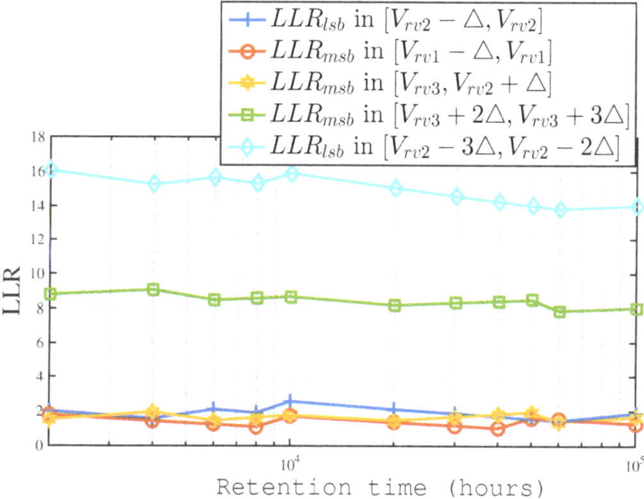

Figure 4. Plot of the relative log-likelihood ratio (LLR) versus data retention time at $PE = 1K$, $\triangle = 0.05$ and $s = 1$.

3.2. Why Are Artificial Neural Networks Useful for NAND Flash Memory?

To simplify the analysis, this subsection first discusses the case that the CCI is only generated by the vertical neighboring cell. In this case, the conditional probability distribution function of the threshold voltage, (8), is simplified to (16):

$$\begin{aligned} p(V_{th}|k \in \{11, 01, 00, 01\}) = \frac{1}{4}[\mathcal{N}(\mu_k - \mu_t, \sigma_k^2 + \sigma_t^2 + \sigma_r^2) \\ + \sum_{\mu_p} \mathcal{N}(\mu_k + \gamma_y(\mu_p - \mu_e) - \mu_t, {\sigma_k}^2 + \gamma_y^2(\sigma_p^2 + \sigma_e^2 \\ + 2\sigma_r^2) + \sigma_t^2 + \sigma_r^2)]. \end{aligned} \tag{16}$$

In (16), it is seen that the threshold voltage distribution can be divided into four parts: the distribution of cells with CCI from "11"-state, "01"-state, "00"-state and "10"-state, which are also shown in Figure 4. In an overlap region, the bits with different CCI noise levels may have different error rates. For instance, in the overlap region between "01"-state and "00"-state, the bits of the cells in "00"-state with CCI from neighboring cells in "11"-state may be wrongly detected as "1" in LSB. In general, we want to find the optimal reading reference voltage at the intersecting point of the distributions of two states, such as the red dotted line in Figure 5. However, once we know the programmed state or the threshold voltage of the cells that donate the CCI to victim cells, the optimal reading reference voltage may change. For example, the optimal reading reference voltage should be selected by the blue dotted line in Figure 5, when the vertical neighboring cell is in the erased state.

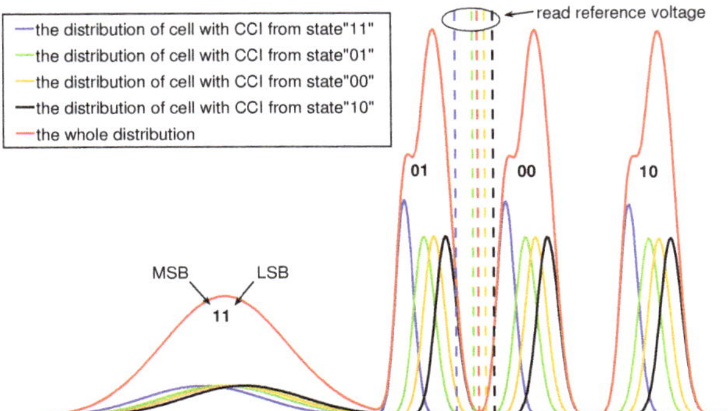

Figure 5. Illustration of the distribution of NAND flash memory at $s = 1.4$ (the cell-to-cell coupling strength factor), $PE = 1K$ and $Retention\ time = 10^5$.

In this paper, we expand the two-dimensional coordinates to three-dimensional, as shown in Figure 6a. The X-axis is the victim cell's voltage, and the Y-axis is the threshold voltage of vertical neighboring cell. By doing so, one can easily find the incorrectly detected cells, marked with red dots. Moreover, we have two important observations:

1. The correct cells (the blue dots) and the incorrect cells (the red dots) are not interlaced in the three-dimensional space. It means that the correct cells (or the incorrect cells) have similar features, which may be used for clustering them from the incorrect ones.
2. The hard decision may not be the optimal decision when the surrounding cells have been read. In Figure 6a, the gray plane is the hard-decision plane, but not optimal. Suppose that there is a decision plane, shown as Figure 6b, and then we apply this plane to the same data in Figure 6a. One can see that the decision performance by the plane gets significantly improved compared to the plane in Figure 6a.

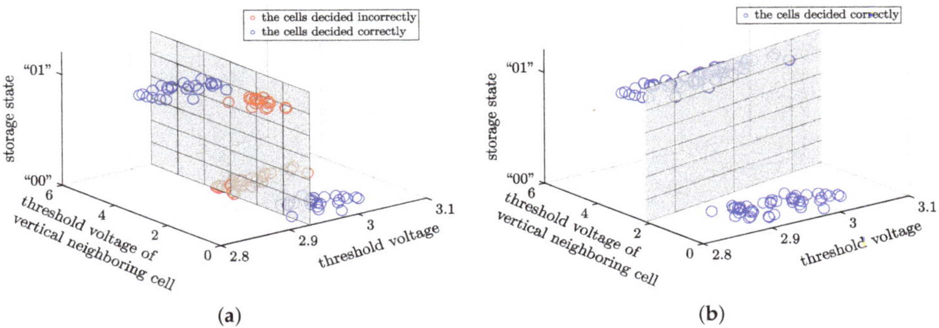

Figure 6. Illustration of the decision of least significant bit (LSB) in the NAND flash memory. (**a**) The conventional hard-decision plane in the three-dimensional coordinates. (**b**) The optimal plane.

These two observations reveal that the detection of bits in a cell can be transformed into a clustering problem, which is to obtain an optimal classification hyperplane. When more surrounding cells are considered, the clustering problem will become more complex and the dimensions of the classification hyperplane will increase beyond three. To address this issue, We propose to use the neural network, which is good at solving various clustering problems.

3.3. Proposed Artificial Neural Network-Assisted Error Correction (ANNAEC) Scheme

The main idea of the proposed ANNAEC scheme is shown in Figure 7. In general, the flash memory controller uses soft-decision error correction [12], read-retry [1,16] and voltage optimization, which has been widely used in practical systems, to ensure the reliability of data stored in NAND flash memory. When these techniques are not effective in suppressing flash channel noise, the flash memory controller attempts to operate the proposed ANNAEC scheme to correct error bits. Moreover, it can reduce the power consumption and computation burden of the controller, since the cells in an overlap region take a relatively small part of the cells on a page.

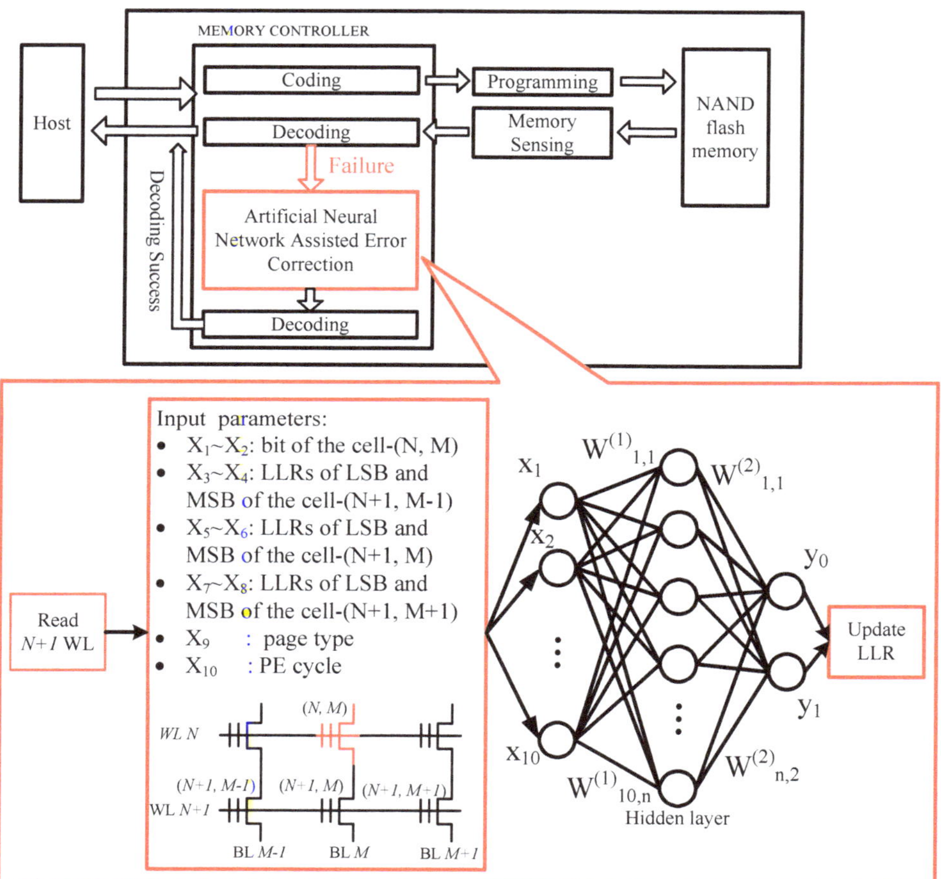

Figure 7. Block diagram of the proposed ANNAEC scheme in NAND flash memory.

In general, the host implements data writing and reading to the NAND flash memory chip by communicating with the memory controller, which communicates with the NAND flash memory chip. First, the host transfers data to the flash controller. The flash controller then encodes the data and writes it into the NAND flash memory chip. When the host reads the data, the flash controller communicates with the NAND flash chip. During this process, the NAND flash chip reads the data from the cell and sends it to the flash controller by reading the sensing circuit. After that, the flash controller corrects and restores the original data through the decoding algorithm and sends it to the host. The proposed a neural network assisted error correction algorithm is used as an alternative decoding algorithm. When the decoding of the flash controller fails, the neural network model is used to first correct the data and then perform decoding.

We label the positions of the cells in an overlap region, which is at the N-th word-line and the M-th bit-line in the block as (N, M), shown in Figure 7. The input parameters of the neural network are summarized in Table 1. X_1 and X_2 are the bits of cell-(N, M) in MLC memory, respectively. $X_3 \sim X_8$ are the LLRs of LSB and MSB of the immediate adjacent cells, i.e., cell-$(N+1, M-1)$, cell-$(N+1, M)$ and cell-$(N+1, M+1)$. X_9 is the flag of page type. If the current reading page is LSB, we set X_9 to "0"; otherwise, X_9 is set to "1". X_{10} is the number of PE cycles. There are two reasons for choosing those parameters: (1) the threshold voltage is difficult to be obtained in a practical system, but the LLR and bits in a cell can help to locate the range of threshold voltage; (2) the vertical and the diagonal neighboring cells contribute about 81% of the CCI [17,18].

Table 1. Summary of input parameters.

Notation	Physical Meaning
X_1, X_2	bit of the cell (N, M)
X_3, X_4	LLRs of LSB and MSB of the cell-$(N+1, M-1)$
X_5, X_6	LLRs of LSB and MSB of the cell-$(N+1, M)$
X_7, X_8	LLRs of LSB and MSB of the cell-$(N+1, M+1)$
X_9	page type (LSB:0; MSB:1)
X_{10}	PE cycle

Afterward, we send the parameters into the back propagation neural network to correct error bits. The sigmoid function is selected as the activation function of the back propagation neural network, given as

$$f(x) = \frac{1}{1+e^{-x}}. \tag{17}$$

The cost function is chosen as the typical mean square error (MSE) cost function [19], given by

$$E = \frac{1}{2}[(T_{y_0} - y_0)^2 + (T_{y_1} - y_1)^2], \tag{18}$$

where the outputs of neural networks y_0 and y_1 are the reliabilities of "0" and "1", and T denotes the desired reliability in the data set. The relative LLR is calculated offline in the flash memory controller. It is difficult to recalculate the relative LLR, since the online characteristic estimation of the memory channel causes longer read latency. Since the accurate relative LLR is hard to recalculate, we update relative LLR by

$$LLR_{update} = (-1)^{\varepsilon+1} \left| LLR_{original} \right|, \tag{19}$$

where $LLR_{original}$ denotes original relative LLR obtained in the sensing operation, and ε is given by

$$\varepsilon = \begin{cases} 1 & \text{if } y_1 > y_0 \\ 0 & \text{else.} \end{cases} \tag{20}$$

Although (19) does not update the accurate LLR to decode, it can estimate the value of LLR. Moreover, (19) is used to correct the sign of LLR, which is more important than the absolute value of LLR, since fewer error signs of LLRs fewer less error bits.

4. Experiment Results

4.1. Training

Throughout all experiments, we used a rate-0.9 (4544, 4096) QC-LDPC code and the BP decoding algorithm. The experimental platform is implemented in Matlab. The channel parameters, which are used to generate the training dataset, are shown in Table 2. Since the parasitic coupling capacitances of CCI are invariable in a flash memory ship, without loss of generality, we set the cell-to-cell coupling strength factor to be $s = 1$. According to the raw bit error rate (RBER), we generate the dataset at $PE = \{3000, 4000, 5000\}$ and divide the dataset into two parts: error and correct bits, which are to be corrected, e.g., the cell-(N, M) in Figure 7. In total, the sizes of the training and validation data are 336,000 and 84,000, respectively. According to the performance of neural network versus the different numbers of hidden layer node, shown in Figure 8, the basic neural network structure is set to be $\{10, 3, 2\}$, meaning that there are 10 nodes in the input layer, 3 nodes in the hidden layer and 2 nodes in the output layer.

Table 2. Training dataset ($s = 1$).

Retention Time (h) / *PE* / RBER	3000	4000	5000
$\approx 6 \times 10^{-3}$	1×10^5	2×10^4	1×10^4
$\approx 7 \times 10^{-3}$	2×10^5	4×10^4	1.5×10^4
$\approx 8 \times 10^{-3}$	3×10^5	5×10^4	2×10^4
$\approx 9 \times 10^{-3}$	5×10^5	1×10^5	3×10^4
$\approx 1 \times 10^{-2}$	1×10^6	5×10^5	1×10^5
Size of the training data	336,000		
Size of the validation data	84,000		

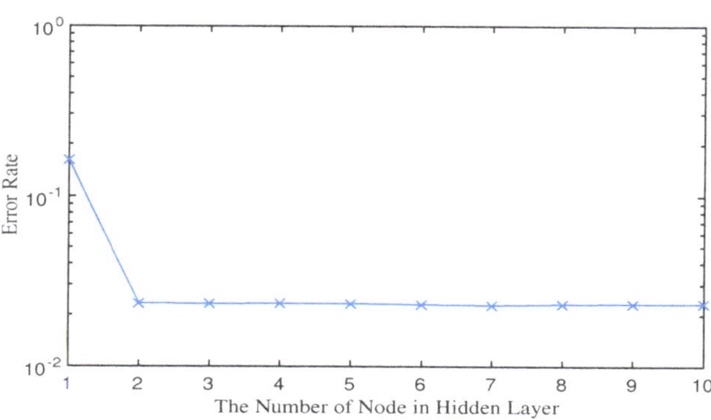

Figure 8. Performance of neural network under the different numbers of hidden layer nodes.

4.2. Performance

In Figure 9a,b, we compare RBER and frame error rate (FER) using ANN-LDPC [11], the proposed method and the original method without the neural network versus data retention time at $s = 1$. We can observe that the proposed ANNAEC significantly reduces the RBER in comparison with the ANN-LDPC and original method.

For instance, in Figure 9a, the data retention time is about 3×10^4 h at $PE = 5000$ and RBER $= 2 \times 10^{-2}$, using the scheme without ANNAEC. Compared to the proposed ANNAEC scheme, Figure 9b shows that for the same performance, the ANN-LDPC can make the flash memory endure up to 3×10^5 h and the proposed method provides a performance gain of approximately 67% of data retention, which makes the retention time of flash endure up to 5×10^5 h. In addition, the proposed method has a more stable error correction performance, when the memory suffers from a weak interference. Similarly, we can notice that the proposed ANNAEC improves the FER performance by up to an error rate of 1×10^{-3} at a retention time of 4×10^6 h and $PE = 3000$. The ANN-LDPC has a FER performance of approximately 5×10^{-3}.

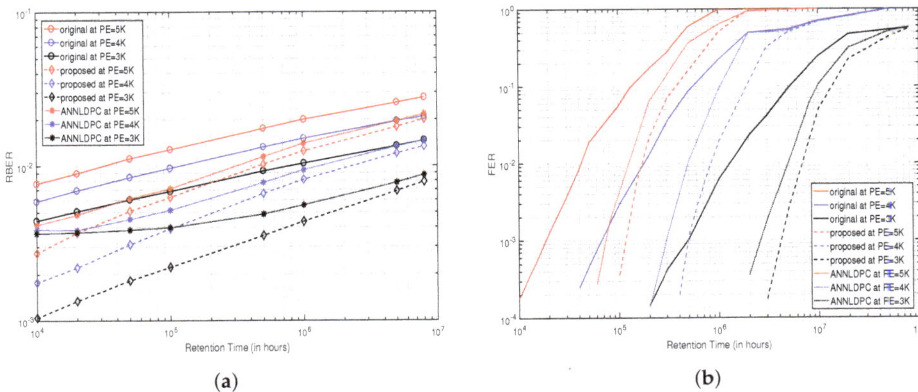

Figure 9. (**a**) Comparison of the raw bit error rate (RBER) performance of NAND flash memory with and without ANNAEC scheme versus data retention time at $s = 1$. (**b**) Comparison of the frame error rate (FER) performance of low-density parity-check (LDPC) coded NAND flash memory with and without the ANNAEC scheme versus data retention time at $s = 1$.

5. Conclusions

In this paper, we have proposed to use the relative LLR calculation to estimate the actual LLR. Furthermore, in three-dimensional coordinates, we have transformed the bit detection problem into a clustering problem, which allows us to apply an artificial neural network in the memory channel. To solve the clustering problem, we proposed an artificial neural network-assisted error correction scheme, which has been shown by experiments to be effective in correcting the error bit when the conventional method without the neural network fails to decode. Simulation results have shown that the FER performance of our ANNAEC is significantly better than that of ANN-LDPC. For example, the ANN-LDPC can make the flash memory endure up to 3×10^5 h, and the proposed method provides the performance gain of approximately 67% of data retention, which makes the retention time of flash endure up to 5×10^5 h. Furthermore, our proposed approach can be extended to TLC or QLC flash memories.

Author Contributions: Conceptualization, R.H., H.H. and G.H.; methodology, R.H. and G.H.; software, R.H.; validation, R.H. and H.H.; formal analysis, R.H., H.H. and G.H.; investigation, R.H., H.H., G.H. and C.X; writing—original draft preparation, R.H., G.H. and C.X.; writing—review and editing, H.H. and G.H.; visualization, C.X.; supervision, G.H.; project administration, R.H. All authors have read and agreed to the published version of the manuscript.

Funding: This research was funded by the National Natural Science Foundation of China under grant 61871136.

Data Availability Statement: The study did not report any data.

Conflicts of Interest: The authors declare no conflict of interest.

References

1. Cai, Y.; Ghose, S.; Haratsch, E.F.; Luo, Y.; Mutlu, O. Error Characterization, Mitigation, and Recovery in Flash-Memory-Based Solid-State Drives. *Proc. IEEE* **2017**, *105*, 1666–1704. [CrossRef] [CrossRef]
2. Lee, J.D.; Choi, J.H.; Park, D.; Kim, K. Data retention characteristics of sub-100 nm NAND flash memory cells. *IEEE Electron Device Lett.* **2003**, *24*, 748–750. [CrossRef]
3. Peng, Z.; He, R.; Han, G.; Cai, G.; Fang, Y. Neighbor-A-Posteriori Information Assisted Cell-State Adaptive Detector for NAND Flash Memory. *IEEE Commun. Lett.* **2019**, *23*, 1967–1971. [CrossRef] [CrossRef]
4. Xiong, Q.; Wu, F.; Lu, Z.; Zhu, Y.; Zhou, Y.; Chu, Y.; Xie, C.; Huang, P. Characterizing 3D Floating Gate NAND Flash. *ACM Trans. Storage* **2018**, *14*, 1–31. [CrossRef] [CrossRef]
5. Dong, G.; Li, S.; Zhang, T. Using Data Postcompensation and Predistortion to Tolerate Cell-to-Cell Interference in MLC nand Flash Memory. *IEEE Trans. Circuits Syst. I Regul. Pap.* **2010**, *57*, 2718–2728. [CrossRef] [CrossRef]
6. Adnan Aslam, C.; Guan, Y.L.; Cai, K. Detector for MLC NAND Flash Memory Using Neighbor A-Priori Information. *IEEE Trans. Very Large Scale Integr. (VLSI) Syst.* **2016**, *24*, 2827–2836. [CrossRef] [CrossRef]
7. Lin, X.; Han, G.; Ouyang, S.; Li, Y.; Fang, Y. Low-complexity detection and decoding scheme for LDPC-coded MLC NAND flash memory. *China Commun.* **2018**, *15*, 58–67. [CrossRef] [CrossRef]
8. Aslam, C.A.; Guan, Y.L.; Cai, K. Decision-Directed Retention-Failure Recovery With Channel Update for MLC NAND Flash Memory. *IEEE Trans. Circuits Syst. I Regul. Pap.* **2018**, *65*, 353–365. [CrossRef] [CrossRef]
9. Riaz, H.; Park, J.; Choi, H.; Kim, H.; Kim, J. Deep and Densely Connected Networks for Classification of Diabetic Retinopathy. *Diagnostics* **2020**, *10*, 24. [CrossRef] [CrossRef] [PubMed]
10. Wei, D.; Qiao, L.; Hao, M.; Feng, H.; Peng, X. Reliability prediction model of NAND flash memory based on random forest algorithm. *Microelectron. Reliab.* **2019**, 100–101. [CrossRef] [CrossRef]
11. Nakamura, T.; Deguchi, Y.; Takeuchi, K. Adaptive Artificial Neural Network-Coupled LDPC ECC as Universal Solution for 3-D and 2-D, Charge-Trap and Floating-Gate NAND Flash Memories. *IEEE J. Solid State Circuits* **2019**, *54*, 745–754. [CrossRef] [CrossRef]
12. Dong, G.; Xie, N.; Zhang, T. On the Use of Soft-Decision Error-Correction Codes in nand Flash Memory. *IEEE Trans. Circuits Syst. I Regul. Pap.* **2011**, *58*, 429–439. [CrossRef] [CrossRef]
13. Compagnoni, C.M.; Ghidotti, M.; Lacaita, A.L.; Spinelli, A.S.; Visconti, A. Random Telegraph Noise Effect on the Programmed Threshold-Voltage Distribution of Flash Memories. *IEEE Electron Device Lett.* **2009**, *30*, 984–986. [CrossRef] [CrossRef]
14. Aslam, C.A.; Guan, Y.L.; Cai, K. Read and Write Voltage Signal Optimization for Multi-Level-Cell (MLC) NAND Flash Memory. *IEEE Trans. Commun.* **2016**, *64*, 1613–1623. [CrossRef] [CrossRef]
15. Sandell, M.; Ismail, A. Machine learning for LLR estimation in flash memory with LDPC codes. *IEEE Trans. Circuits Syst. II Express Briefs* **2021**, *68*, 792–796. [CrossRef] [CrossRef]
16. Yong, K.-K.; Chang, L.-P. Error Diluting: Exploiting 3-D NAND Flash Process Variation for Efficient Read on LDPC-Based SSDs. *IEEE Trans. Comput. Aided Des. Integr. Circuits Syst.* **2020**, *39*, 3467–3478. [CrossRef] [CrossRef]
17. Kim, T.; Kong, G.; Weiya, X.; Choi, S. Cell-to-Cell Interference Compensation Schemes Using Reduced Symbol Pattern of Interfering Cells for MLC NAND Flash Memory. *IEEE Trans. Magn.* **2013**, *49*, 2569–2573. [CrossRef] [CrossRef]
18. Park, S.K.; Moon, J. Characterization of Inter-Cell Interference in 3D NAND Flash Memory. *IEEE Trans. Circuits Syst. I Regul. Pap.* **2021**, *68*, 1183–1192. [CrossRef] [CrossRef]
19. Žalik, K.R. An efficient k′-means clustering algorithm. *Pattern Recognit. Lett.* **2008**, *29*, 1385–1391. [CrossRef] [CrossRef]

 *micromachines*

Article

Understanding the Origin of Metal Gate Work Function Shift and Its Impact on Erase Performance in 3D NAND Flash Memories

Sivaramakrishnan Ramesh *, Arjun Ajaykumar, Lars-Åke Ragnarsson, Laurent Breuil, Gabriel Khalil El Hajjam, Ben Kaczer, Attilio Belmonte, Laura Nyns, Jean-Philippe Soulié, Geert Van den bosch and Maarten Rosmeulen

IMEC, Kapeldreef 75, B-3001 Leuven, Belgium

* Correspondence: siva.ramesh@imec.be

Abstract: We studied the metal gate work function of different metal electrode and high-k dielectric combinations by monitoring the flat band voltage shift with dielectric thicknesses using capacitance–voltage measurements. We investigated the impact of different thermal treatments on the work function and linked any shift in the work function, leading to an effective work function, to the dipole formation at the metal/high-k and/or high-k/SiO_2 interface. We corroborated the findings with the erase performance of metal/high-k/ONO/Si (MHONOS) capacitors that are identical to the gate stack in three-dimensional (3D) NAND flash. We demonstrate that though the work function extraction is convoluted by the dipole formation, the erase performance is not significantly affected by it.

Keywords: work function; effective work function; dipole; metal gate; high-k; SiO_2; interfacial reaction; MHONOS; erase performance; 3D NAND flash memory

Citation: Ramesh, S.; Ajaykumar, A.; Ragnarsson, L.-Å.; Breuil, L.; El Hajjam, G.K.; Kaczer, B.; Belmonte, A.; Nyns, L.; Soulié, J.-P.; Van den bosch, G.; et al. Understanding the Origin of Metal Gate Work Function Shift and Its Impact on Erase Performance in 3D NAND Flash Memories. *Micromachines* **2021**, *12*, 1084. https://doi.org/10.3390/mi12091084

Academic Editors: Cristian Zambelli and Rino Micheloni

Received: 4 August 2021
Accepted: 1 September 2021
Published: 8 September 2021

1. Introduction

When it comes to low-cost and large density non-volatile memory, three-dimensional (3D) NAND flash memory technology is the industry standard [1,2]. The memory stack used in 3D NAND is inspired by a typical SONOS memory cell, which allows easy vertical integration and is addressed by horizontal word lines (WL). To improve the bit density, the number of cells in the vertical 3D NAND string is increased. This requires the stacking of many WLs, which need to be as thin as possible to limit the total height and mechanical stress of the structure [3]. Tungsten (W) metal-based WL is currently being used by the industry. However, novel materials with lower resistivity are being considered as future candidates to reduce the high resistive-capacitive (RC) delay that results as a consequence of WL thinning and continued stacking of the WLs (i.e., downscaling the metal thickness) in the vertical direction.

Moreover, the WL metal can act as an enabler to improve the 3D NAND erase operation. It was shown that high work function metals, such as TiN and Ru, can delay the electron injection from the gate (i.e., electrons tunneling from the gate into the charge-trap layer), thereby improving the erase window [4]. It has also been demonstrated [5] that when a metal gate is used in combination with a thin high-k liner, such as Al_2O_3, HfO_2, or ZrO_2 (i.e., a Metal/High-k/ONO/Si (MHONOS) structure), the erase performance can be further improved. Figure 1 plots the erase saturation levels (lowest possible threshold voltage, V_{TH}, shift achievable) for different scenarios, with and without a high-k liner, as simulated using our in-house developed 1D simulator [6]. The high-k liner helps to lower the injecting field for the electrons at the gate, and even proves to have a larger impact than the metal work function (WF). The erase is found to be penalized when the MHONOS stack is treated with a high thermal budget [3]. To thoroughly investigate the WL metal and high-k liner combination, and its effect on erase operation, metal work function extraction experiments have been proposed and studied in this work.

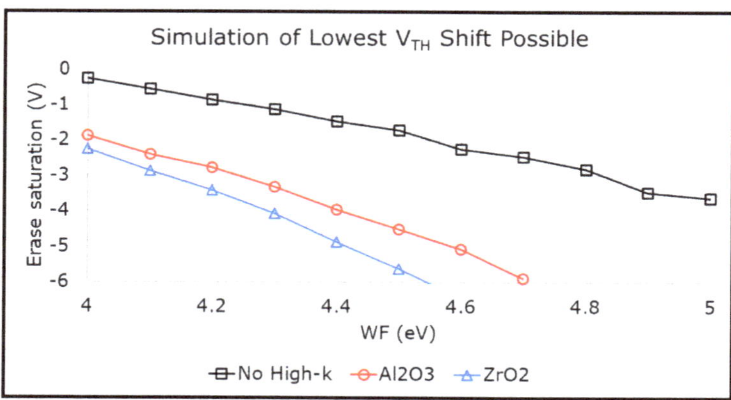

Figure 1. Simulations of erase saturation levels in a memory stack without high-k liner, or with 2 nm Al_2O_3 or ZrO_2. Addition of a high-k liner shows more benefit than (work function) WF.

WF analysis of metal gate electrodes on high-k dielectrics, by monitoring flat-band voltage, V_{FB} (or threshold voltage, V_{TH}), have been demonstrated in the literature [7–12]. The studies report an undesirable shift in the V_{FB} (or V_{TH}) of metal-oxide-semiconductor (MOS) devices. The origins are unclear, leading to an effective work function (eWF) for the metal, different from the bulk values. Some reports in the literature attribute this shift to Fermi level pinning (FLP) caused either by metal-induced gap states [13–15] or charged defects/oxygen transfers, at the metal/high-k interface [12,16,17]. Dipole formation at the high-k/SiO_2 interface due to oxygen vacancies [18,19], and/or the energy offsets between the high-k and SiO_2 [20], have also been suggested in the literature as possible root causes for an eWF. Though, these studies suggest a notable dependence of eWF on the choice of high-k used, other process parameters such as gate electrode deposition and annealing conditions have been found to affect the eWF in a significant way as well [21].

In this paper, we investigate the change in WF (i.e., eWF) of metal electrodes deposited on high-k dielectrics. Based on the process conditions used, we evidence it to either the interfacial reactions at the WL-to-high-k contact or between the high-k and the oxide. The aim of this work is to understand the origins and consequences of WF shifts based on process conditions within the context of 3D NAND flash memory devices. Therefore, we also analyze various MHONOS stacks containing Al_2O_3, ZrO_2, HfO_2 high-k liners and TiN, Ru, Mo as gate metal, and corroborate the eWF with the erase performance of these stacks.

2. Materials and Methods

Capacitors with and without the charge trap layer were fabricated on 300 mm p-doped Si (100) wafers for erase analysis and WF extraction, respectively.

2.1. Work Function Extraction Methodology

The WF of a metal on high-k is determined by extracting V_{FB} from capacitance–voltage (CV) measurements on a metal-insulator-semiconductor (MIS) structure [22]. The schematic in Figure 2 shows the energy band diagram of an MIS structure. From this, we note that the metal work function can be expressed as follows

$$\Phi_M = V_{FB} + \chi_{Si} + [E_C - E_F], \tag{1}$$

where Φ_M is metal work function, V_{FB} is flat-band voltage computed from CV measurements, χ_{Si} is electron affinity of Si substrate, E_C and E_F are the conduction band minima and fermi level.

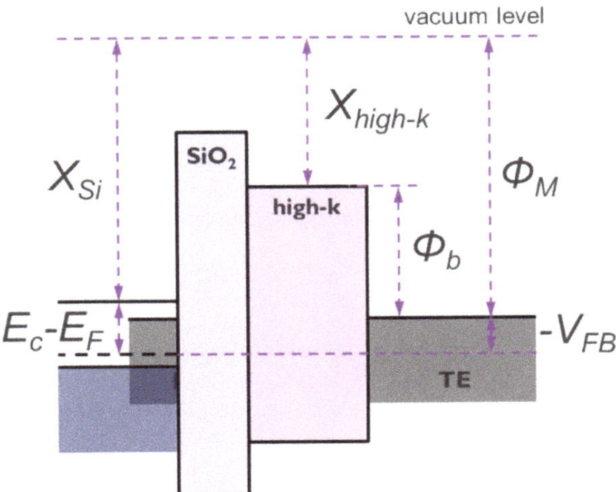

Figure 2. Schematic of the energy band diagram of a metal-insulator-semiconductor (MIS) capacitor.

However, the charges present in the bulk and at the interfaces of the oxides [23] can affect the V_{FB} as follows

$$\Delta V_{FB} = \int_0^{t_{ox}} \frac{\rho(z)(t_{ox} - z)}{\varepsilon(z)\varepsilon_0} dz, \tag{2}$$

From the above equation, it is clear that the effect of these oxide charges can be cancelled out by extracting the V_{FB} at zero oxide thickness. This calls for variations in SiO_2 and high-k thicknesses. With the help of a slant etch technique, the thickness of SiO_2 was varied across the wafer as shown in schematic in Figure 3. For each electrode, a set of 3 wafers with different high-k thicknesses (typically 3 nm, 5 nm, 7 nm) was used to provide enough variation and extract the WF conveniently. Typical CV measurements and V_{FB} extraction procedure are discussed in Appendix A.

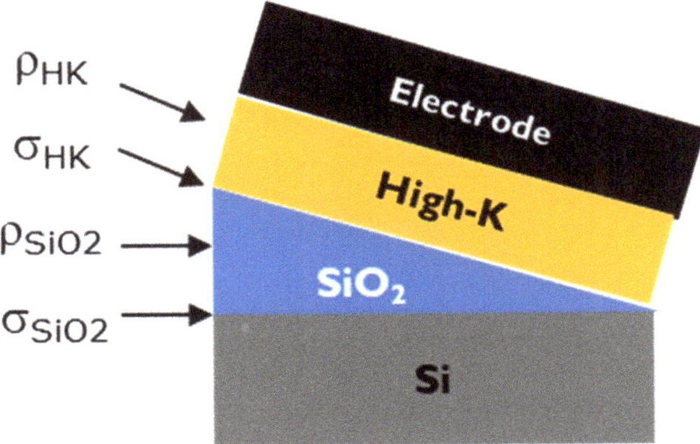

Figure 3. Schematic of MIS capacitor with slant etch for SiO_2. Corresponding oxide charge densities are indicated.

The impact of oxide charges on V_{FB} can be mathematically expressed in terms of equivalent oxide thickness (EOT) and the corresponding charge densities as follows [24]

$$V_{FB} = \Phi_{MS} + q \cdot \rho_{HK} \cdot \varepsilon_{HK} \cdot \frac{EOT_{HK}^2}{2 \cdot \varepsilon_{ox}^2 \cdot \varepsilon_o} + q \cdot \sigma_{HK} \cdot \frac{EOT_{HK}}{\varepsilon_{ox} \cdot \varepsilon_o} + q \cdot \rho_{SiO_2} \cdot \frac{0.5 \cdot T_{SiO_2}^2 + \left(\frac{\varepsilon_{HK}}{\varepsilon_{ox}}\right) \cdot T_{SiO_2} \cdot EOT_{HK}}{\varepsilon_{ox} \cdot \varepsilon_o} + q \cdot \sigma_{SiO_2} \cdot \frac{EOT_{total}}{\varepsilon_{ox} \cdot \varepsilon_o}, \tag{3}$$

where q is the electron charge, ρ_{HK} and σ_{HK} are the bulk and interface charge densities of high-k dielectric, respectively. The terms ρ_{SIO2} and σ_{SIO2} are the corresponding bulk and interface charge densities of SiO_2, respectively. EOT_{HK}, T_{SiO2}, and EOT_{total} are the equivalent oxide thickness of high-k, thickness of SiO_2, and both combined, respectively. The EOT_{total} is in fact the measured EOT computed from the CV measurement of the MIS capacitors. The terms ε_{HK}, ε_{ox}, ε_o are the relative permittivity of high-k, SiO_2 and permittivity of free space, respectively. The Φ_{MS} in the above equation, from which the metal WF is extracted, is later computed by extrapolating V_{FB} at EOT (both high-k and SiO_2) = 0.

First, a 30 nm thick layer of high quality SiO_2 was thermally grown at 900 °C. This was then etched back with a slant profile (as shown in Figure 3) by slowly immersing (at a constant rate) the wafer in a 1.9% hydrofluoric acid (HF) solution. The desired thickness range of SiO_2 is obtained across the wafer by modifying the rate of immersion accordingly. A nominal thickness range of 3–12 nm was used in this work. Then, after the slant etch, a 3 nm plasma enhanced atomic layer deposition (PEALD) SiO_2 was uniformly deposited at 300 °C, to mimic the blocking oxide in a 3D NAND device. Little wafer-to-wafer variations were observed in the oxide thickness, as measured by ellipsometry (see Figure 4a). The total EOT measured from CV will vary across the wafer due to the slant etch of thermal oxide, as shown in Figure 4b (bubble size represents magnitude of EOT).

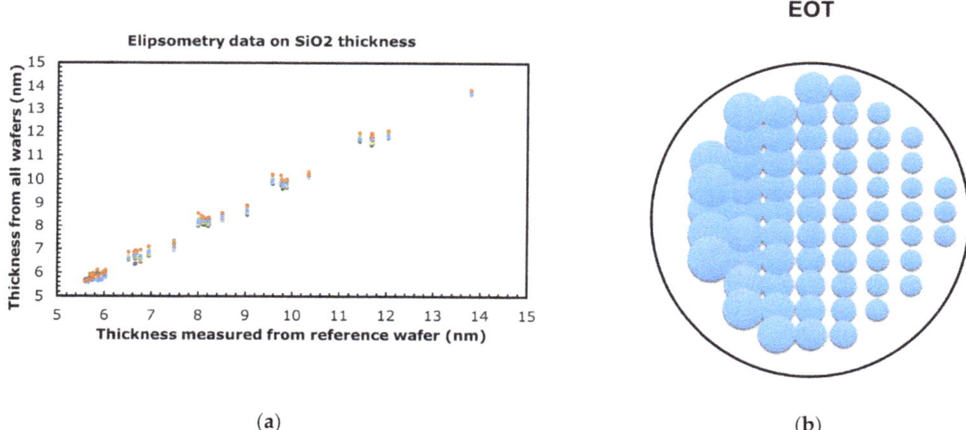

Figure 4. (**a**) Thickness of SiO_2, after slant etch and plasma enhanced atomic layer deposition (PEALD) oxide deposition, measured across multiple wafers using ellipsometry; (**b**) Equivalent oxide thickness (EOT) computed from capacitance–voltage (CV) measurement. Bubble size represents EOT magnitude.

After this, high-k liners, such as Al_2O_3, ZrO_2, and HfO_2, were deposited at 300 °C to their desired thicknesses, using atomic layer deposition (ALD). Finally, 20 nm ALD Ru or ALD TiN or PVD Mo were then deposited as the gate electrode. In order to isolate the impact of thermal treatment on individual layers, a high temperature anneal (T_{anneal}) was performed at different stages of the stack formation (as shown in Figure 5). For instance, some of the capacitors were subjected to a post metallization anneal (PMA) for 20 min at 750 °C in N_2 ambient. A few others were subjected to a post high-k deposition anneal (PDA), where the entire stack sans the metal electrode received a thermal treatment for 1 min at 1050 °C for Al_2O_3-based stacks and 1 min at 750 °C for the rest, all in N_2 ambient. All wafers received a final sintering anneal in 5 atm H_2 ambient at 450 °C for 30 min.

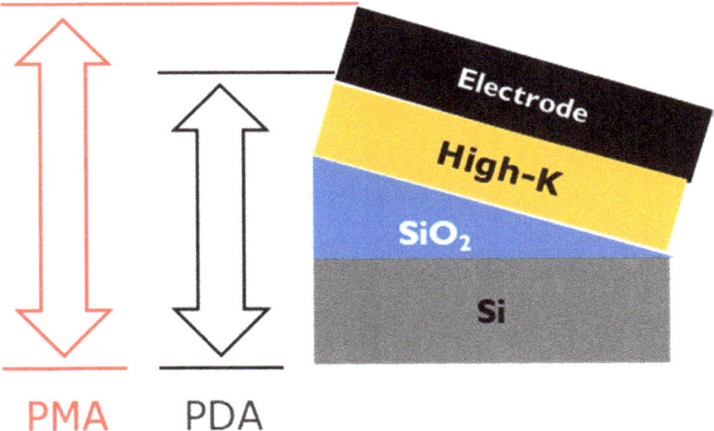

Figure 5. Schematic indicating different anneal types and the corresponding layers that received the process.

CV measurements were performed on 70 × 70 μm^2 capacitors at a frequency of 100 kHz. The parameters needed for the WF extraction, namely, V_{FB}, the substrate doping concentration and the total EOT, EOT_{total}, are estimated (see Appendix A) with the help of NCSU's CVC model fitting software [25]. Based on the expression for V_{FB} from Equation (3), we can express V_{FB} as a second order polynomial equation in terms of the EOT, as the one below

$$V_{FB} = \Phi_{MS} + a \cdot EOT_{HK}^2 + b \cdot EOT_{HK} + p \cdot T_{SiO_2}^2 + q \cdot T_{SiO_2}, \quad (4)$$

where a, b, p, and q contain the charge densities of high-k and SiO_2.

From the above equation, we can first eliminate the effect of charges in SiO_2 with a second order polynomial fit of the V_{FB} with the thickness of SiO_2, T_{SiO2}. A sample fit is shown in Figure 6. The intercept from the first fit contains the polynomial equation with high-k EOT, EOT_{HK} and hence is used to eliminate the charges from high-k in a second fit.

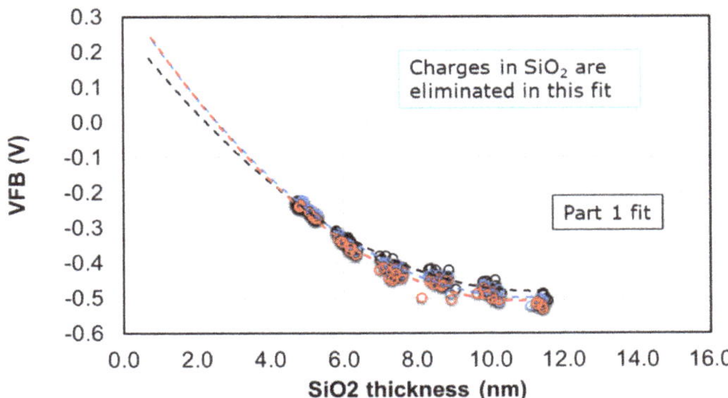

Figure 6. The V_{FB} measured from CV is plotted as a function of SiO_2 thickness. A second order fit is performed to isolate the terms p and q containing the charge densities in its bulk and interface.

As mentioned earlier, we have the EOT_{total} of the stack as measured from CV. In order to get the T_{SiO2} to be used in the first fit, we make use of the ellipsometry data that was measured at preset locations across the wafer, after the slant etch and PEALD deposition. This data is then compared with corresponding dies for which the CV was measured. The difference between the measured EOT_{total} and this ellipsometry data will give an estimate of the EOT_{HK}.

The three curves shown in Figure 6 represent the three wafers with three different high-k thicknesses needed for sufficient variation to eliminate the charges affecting the V_{FB} The corresponding intercept from the 2nd order fit of the above curves is then used in a second fit, as shown in Figure 7 below.

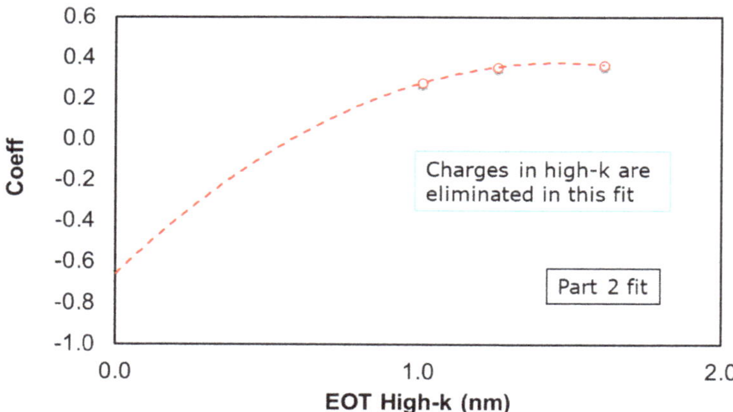

Figure 7. The intercepts from the part 1 fit are plotted as a function of high-k EOT. A second order fit is performed to extract the metal work function.

The intercepts vs. the EOT_{HK} will now help to eliminate the charges in high-k. The intercept from this second fit is the Φ_{MS} from which the WF is computed using the formula

$$WF = 4.05 + \Phi_{MS} + E_C - E_F, \tag{5}$$

where $E_C - E_F(in\ eV) = 1.12 - 0.0257 * \ln\left(\frac{1.83E19}{median\ doping\ concentration\ in\ the\ substrate}\right)$.

2.2. NAND Flash Erase Analysis

Incremental Step Pulse Erase (ISPE) characteristics were studied by monitoring the shift in V_{TH} of MHONOS capacitors from their fresh state. The erase operation is divided into a number of steps with increasing amplitude (for a duration of 1 ms) in applied voltage and at the end of each of them a verify operation is applied to check the V_{TH}. The amplitude and rate of change in V_{TH} is considered as a measure of erase performance.

Large MHONOS capacitors (50 × 50 μm^2) were fabricated on 300 mm p-doped Si (100) wafers, as shown in Figure 8b. N$^+$-doped rings were processed, surrounding the active area of the capacitors, to provide minority carriers for program operation. In a study reported elsewhere [3], we have demonstrated a 3D NAND test structure with 5 layers and showed that the memory characteristics of the stack (see Figure 8a) are qualitatively similar to that of the planar test structures that we typically use (see Figure 8b). Moreover, the gate stack deposited in this work mimics the one of 3D NAND in production [3,26] in terms of annealing processes and high-k/metal gate depositions performed. Therefore, we could fairly say that the results obtained from the planar capacitors in this work are relevant for 3D NAND flash memory devices.

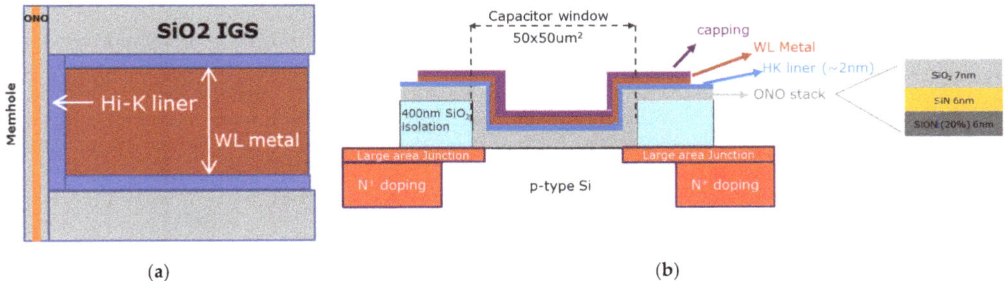

(**a**) (**b**)

Figure 8. (**a**) Cross-section schematic of the memory gate stack in a vertical three-dimensional (3D) NAND device; (**b**) schematic of a planar test structure used in this work. The components of the gate stack are indicated in the figure.

The MHONO stack, as seen from the TEM image in Figure 9a, consists of a 6 nm SiON (with 20% N-to-O ratio) tunnel layer deposited using CVD at 780 °C, 6 nm LPCVD Si_3N_4 charge trap layer deposited at 690 °C, 7 nm PEALD SiO_2 blocking oxide deposited at 300 °C, and 2 nm ALD Al_2O_3 or ZrO_2 or HfO_2 high-k liner deposited at 300 °C. A total of 20 nm ALD Ru or ALD TiN or PVD Mo were then deposited as the gate electrode (WL, wordline). Similar to the study of WF extraction, a post metallization anneal, PMA for 20 min at 750 °C in N_2 ambient, and a post deposition anneal, PDA for 2 min at 1050 °C for Al_2O_3 based stacks and 1 min at 750 °C for the rest, all in N_2 ambient, were performed for some of the capacitors (see Figure 9b). All wafers were subject to a final sintering anneal either in forming gas at 420 °C for 20 min or in 5 atm H_2 ambient at 450 °C for 30 min. We may note that the sintering anneal has little influence on the final erase saturation levels.

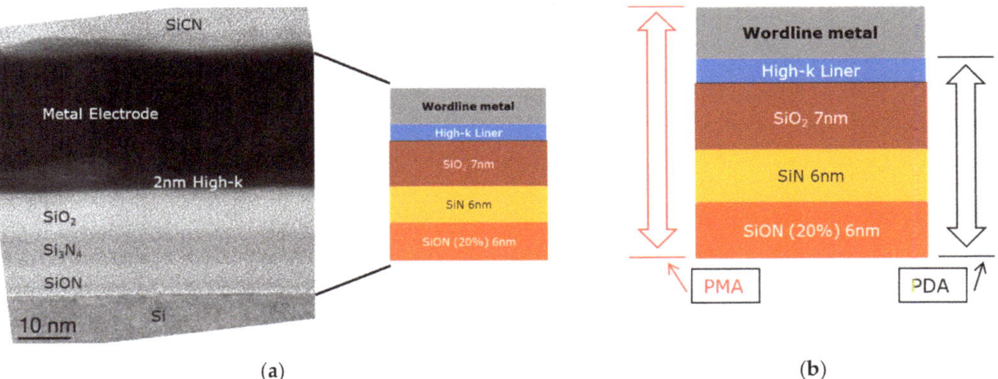

(**a**) (**b**)

Figure 9. (**a**) Transmission electron microscope (TEM) image of a memory stack fabricated in this work; (**b**) different anneal types and the corresponding MHONOS layers that received the anneal.

3. Results and Discussion

The metal WF extracted in this work are listed as a histogram plot in Figure 10 for a few metal/high-k combinations. No high temperature anneals were performed for these splits. W Ref represents the CVD W/thin (3 nm) ALD TiN/Al_2O_3 liner stack similar to the one used currently in 3D NAND production. We could note that the WF of TiN in combination with Al_2O_3 is estimated to be about 4.53 eV and is in close agreement with the actual TiN WF reported in the literature [27,28]. What is surprising is the WF of Ru in combination with Al_2O_3, which is about 200–300 meV less than those reported in the literature for Ru metal [29,30]. It has been demonstrated, using internal photoemission experiments [31], that subtle changes in the chemical bonding at the metal/high-k interface can cause a significant impact on the barrier height (Φ_b, as shown in Figure 2) at this interface. Such chemical modifications could occur from various processing, such as conditions of deposition, thermal budget, and ambient of annealing process. As a consequence, this could lead to a shift in the WF of the metal. However, it is possible to avert this interfacial reaction by using appropriate interfacial layer (IL), as can be seen from Figure 10. The WF of Ru improves to 4.8 eV by adding a thin (3 nm) TiN liner between Ru and Al_2O_3.

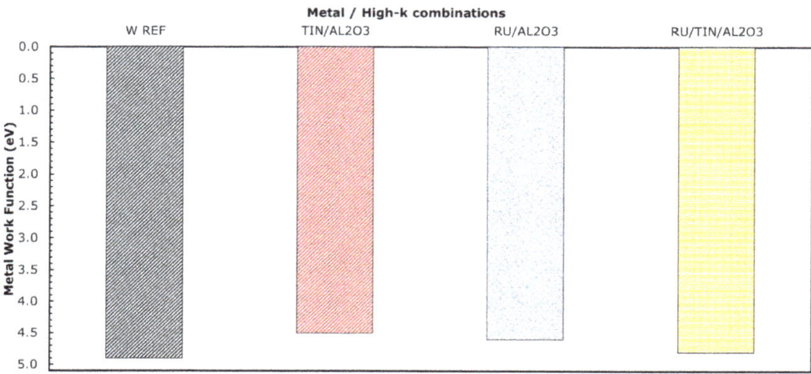

Figure 10. Metal work function listed for a few metal/high-k combinations from this work. No high temperature anneals were performed for these stacks.

In order to verify whether these shifts, measured in WF of Ru, reflect the actual change in metal WF, we compared the erase performance of these stacks. Figure 11 shows the ISPE curves for MHONOS stacks containing the metal/high-k combinations from Figure 10. The erase saturation (lowest VT shift achieved in ISPE) for TiN and Ru on Al_2O_3 (WF ~4.6 eV) is comparable after accounting for the differences in the starting V_{TH}, while that of W Ref (WF ~4.9 eV) is better, corroborating the WF difference between these stacks. With the addition of TiN liner, the WF of Ru improves, and so does the erase saturation.

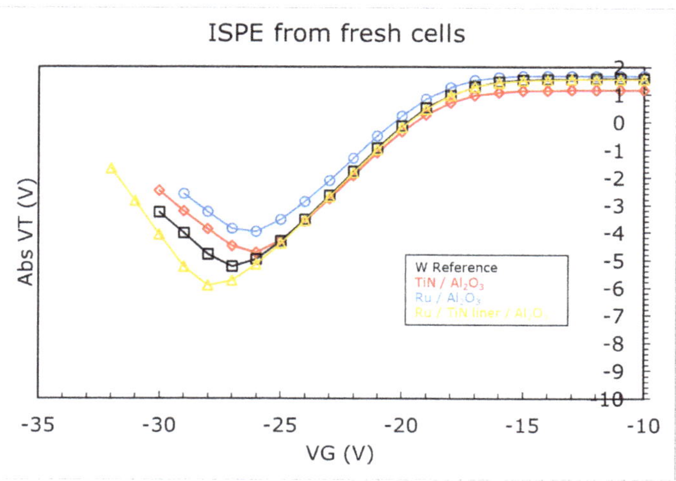

Figure 11. Incremental Step Pulse Erase (ISPE) of (Metal/High-k/ONO/Si) MHONOS, for different metal/high-k combinations from Figure 10.

We may note that the WF extracted from the Ru/TiN/Al_2O_3 stack is slightly less than that of W Ref, i.e., W/TiN/Al_2O_3 stack, yet the erase is better with Ru. Before addressing this, let us look at Figure 12a,b, which display the WF extracted for Ru, Mo, and TiN in combination with HfO_2, ZrO_2, and Al_2O_3 after different annealing conditions, as described in Figure 5. From Figure 12a, we could note a significant reduction (>500 meV) in the WF of Ru after the thermal treatment, irrespective of whether the metal electrode received the anneal (PMA) or not (PDA). The ISPE curves for these stacks are shown in Figure 13a. The stack that received the PDA does not change in erase while the one that received a PMA degrades both in erase slope and saturation level. We can also note from Figures 12b and 13b that without any high temperature anneals, both Ru and Mo show similar WF and erase saturation levels in combination with ZrO_2. Though after a thermal treatment (PMA or PDA), the WF reduces irrespective of the metal or high-k used, the erase saturation depends on the type of anneal applied. These observations (made from Figure 10, Figure 12, Figure 13) hint that (a) the WF alone is not the reason for erase functionality, and (b) an extra factor, unaccounted in the extraction, is affecting the WF, resulting in an effective work function, eWF, being measured from the experiments.

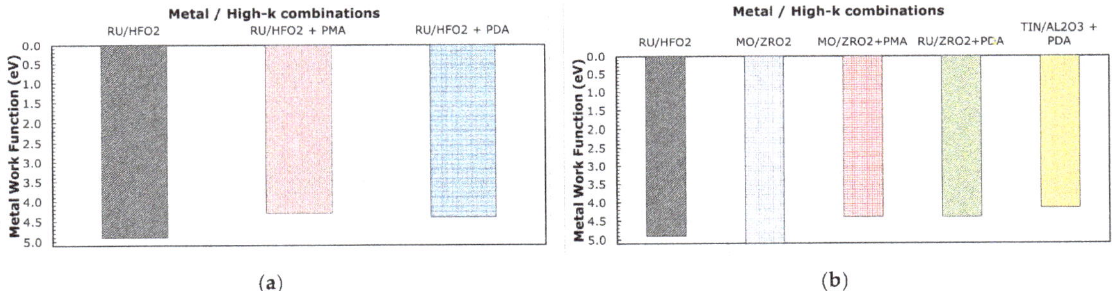

Figure 12. WF extracted for multiple metal and high-k combinations after different annealing conditions. (**a**) Ru with HfO_2; (**b**) Ru, Mo with ZrO_2, and TiN with Al_2O_3.

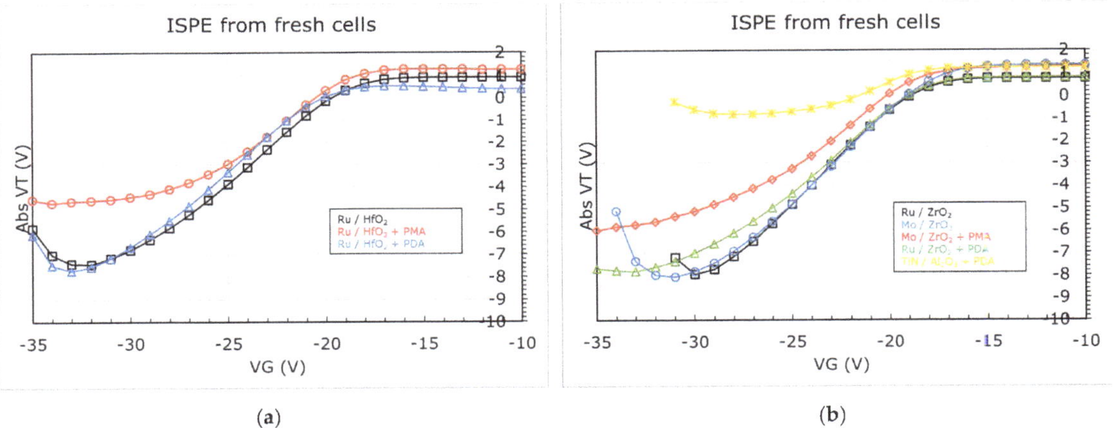

Figure 13. ISPE of MHONOS stacks for (**a**) Ru/HfO_2. Erase performance degrades with post metallization anneal (PMA) while no change after a post high-k deposition anneal (PDA); (**b**) Ru and Mo with ZrO_2 and TiN with Al_2O_3. Similar degradation after PMA as in the case with HfO_2. However, worse performance with Al_2O_3.

It is important to note that in the case of TiN with Al_2O_3 (PDA performed at 2min 1050 °C), the degradation in erase saturation is much worse, which is unlike the observations made for HfO_2- and ZrO_2-based stacks, and definitely not reflected in the WF reduction in TiN. A closer study on the high-k material properties reported elsewhere [32], investigated by trap spectroscopy, revealed that worse erase saturation levels at increased thermal budgets could be due to an increase in defect density in the high-k rather than a reduction in the metal WF itself. Higher defect density could increase trap-assisted tunneling [33], thereby increasing the leakage current during the erase operation (a typical band diagram during erase can be seen in Figure 14).

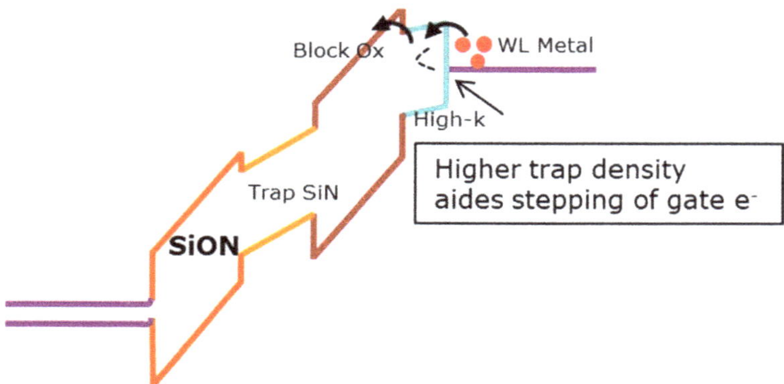

Figure 14. A typical band diagram of MHONOS during erase. Higher trap density reduces the tunneling path for gate electrons resulting in poor erase.

As discussed before, Fermi level pinning (FLP) at the metal/high-k interface, dipole formation at the high-k/SiO_2 interface, and/or the energy offsets between the high-k and SiO_2 have been suggested in the literature as possible root causes for an eWF. If the metal fermi level is pinned, then the Φ_b at the interface should be different, which reflects in the erase saturation levels. Based on the observations made from Figure 11 for Ru with TiN liner and Figure 13 for Ru stacks after PDA, this effect can be ruled out. A common opinion in the literature [21,34–36] is that a dipole formed at the high-k/SiO_2 interface is the dominant factor causing appreciable shifts in V_{FB}, and hence, the WF extracted from it. Many physical models exist to explain this dipole formation, attributing it to dielectric contact induced gap states [37] or dictated by the electronegativity and ionic radii of the cations (from the high-k) [38], However, the most acceptable explanation seems to be oxygen vacancies driven by structural stabilization at the high-k/SiO_2 interface [18–21,34,39,40]. Moreover, the dipole formation at the high-k/SiO_2 interface should not affect the erase performance of flash memory, which is determined by the electron injection dynamics at the gate contact.

To further clarify the impact of dipole formation on erase performance of flash memory, dipole-forming interlayers (DIL) [36,41,42], namely, Al_2O_3 and La_2O_3 (0.6 nm each), were studied as part of the MHONOS stack (shown in Figure 15). The DIL were deposited between metal and high-k or high-k and SiO_2, with TiN/HfO_2 being used as the control gate electrode and high-k dielectric. All the stacks received a PDA for 1.5 s at 1050 °C in N_2 ambient. The corresponding shifts in V_{FB} caused by the interlayers were extracted from CV measurements using CVC fitting (as can be seen in Figure 16).

Figure 15. Schematic of MHONOS stacks with dipole-forming interlayers at different locations. Al_2O_3 and La_2O_3, each 0.6 nm thin, were used as interlayers with HfO_2 used for high-k value.

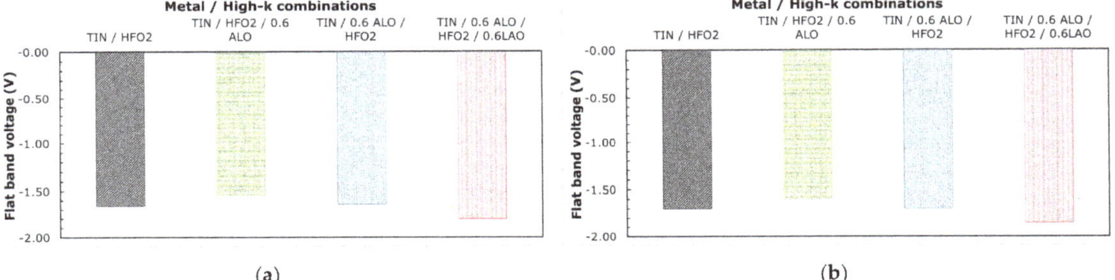

Figure 16. Flat band voltage monitored from CV traces, for MHONOS stacks with different dipole interlayers from Figure 15 (**a**) without any PMA and (**b**) with PMA for 20 min at 750 °C in N_2 ambient.

We could note from Figure 16a that with the addition of Al_2O_3 DIL between HfO_2 and SiO_2, the V_{FB} positively increases by about 120 meV, while it remains unchanged when Al_2O_3 is inserted between the metal and high-k. Though much higher V_{FB} shifts are theoretically reported for Al_2O_3 [18], the processing conditions and thickness of the DIL play a major role in determining the magnitude of the V_{FB} shifts [21,42,43]. Furthermore, if we add 0.6 nm La_2O_3 DIL between HfO_2 and SiO_2 while keeping the Al_2O_3 between TiN and HfO_2, we notice a negative drop of about 140 meV in the V_{FB}, which is in line with trends reported in the literature [44,45]. It is worth to note that the trend in V_{FB} remains unchanged after a PMA for 20 min at 750 °C in N_2 ambient (see Figure 16b).

The ISPE curves for these stacks without PMA are shown in Figure 17a. We could note, despite the differences in V_{FB}, that there is no difference in the erase performance of these stacks. On the contrary, when the stacks were subjected to PMA, the erase depends on the material present in the stack, as can be seen in Figure 17b. The control sample with only TiN and HfO_2 shows slight degradation after PMA. However, the stacks with DIL show higher reduction in erase, even worse when the Al_2O_3 is present next to the blocking oxide, though it shows a positive V_{FB} shift (indicating a higher eWF). It is well known that Al_2O_3 dielectric suffers from a wider band of defect profile [46]. Recalling the discussion from before on the possible impact of defect density in the high-k on erase (see Figure 14), we could fairly say that the above results corroborate this hypothesis.

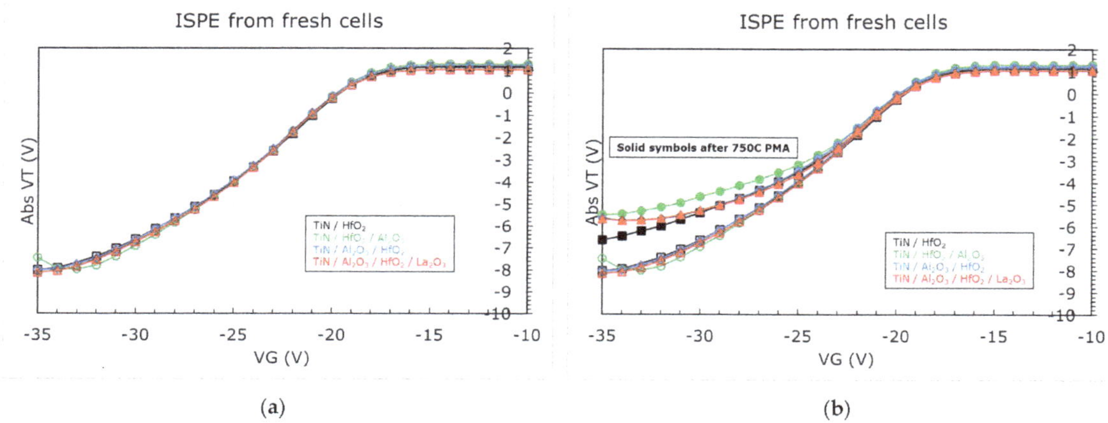

Figure 17. ISPE of MHONOS, for metal/high-k combinations from Figure 16. (**a**) Without any PMA; (**b**) with PMA for 20 min at 750 °C in N_2 ambient.

4. Conclusions

We have extracted and studied the shifts in metal work function (i.e., effective work function, eWF), in response to different processing parameters, such as gate electrode and high-k dielectric materials, and variations in annealing conditions. By studying the work function in combination with the erase performance of NAND flash memory, we were able to narrow down the origin of eWF to dipole formation due to (a) interfacial reactions at the metal/high-k interface and/or (b) possible oxygen vacancies driven by structural stabilization at the high-k/SiO_2 interface. It must be noted that based on the above studies, we did not observe fermi level pinning at the metal/high-k interface.

We also verified and validated the negligible impact of dipole on erase performance by studying different dipole forming interlayers in the memory cell. It is clear that the metal WF extraction is convoluted by dipole formation, while the erase performance of a flash memory cell is affected more by the trap profile in the high-k liner than any other factors that cause shift in flat band voltage.

Author Contributions: S.R. was involved in the conceptualization, investigation, formal analysis, visualization, and data curation. The author was also involved in writing the original draft of this manuscript. A.A. was involved in the conceptualization and investigation. L.-Å.R. was involved in the conceptualization, supervision, and validation. L.B. was involved in the conceptualization, investigation, validation, and writing—review and editing. The author was also involved in formal analysis and visualization in the early stages of this work. G.K.E.H. was involved in the conceptualization and investigation in the early stages of this work. B.K. and A.B. were involved in the methodology and software development required for the formal analysis in this work. L.N. and J.-P.S. were involved in the investigation and developing material resources for this work. G.V.d.b. and M.R. were involved in conceptualization, project administration, supervision, and writing—review and editing. All authors have read and agreed to the published version of the manuscript.

Funding: This work is supported by IMEC's Industrial Affiliation Program on Storage Memory devices.

Data Availability Statement: Not applicable.

Acknowledgments: Authors acknowledge IMEC's pilot line, the various engineering, process, and support teams. Authors also acknowledge HPSP Inc. for the high-pressure annealing experiment.

Conflicts of Interest: The authors declare no conflict of interest.

Appendix A

Figure A1 shows the typical capacitance and conductance curves obtained on 70 × 70 μm^2 capacitors at a frequency of 100 kHz. The capacitors were fabricated on a p-doped, 300 mm Si substrate with SiO_2 bevel. Data is shown for 3 nm HfO_2 high-k liner and Ru gate electrode. The capacitors are sequentially measured at different voltage sweep ranges (i) 1 V to −1 V, (ii) 2 V to −2 V, (iii) 3 V to −3 V. We could notice that there is little impact of the voltage sweep on the hysteresis of the curves. Capacitance data from the 2 V to −2 V voltage sweep range is then used to subsequently extract the flat band voltage, V_{FB}.

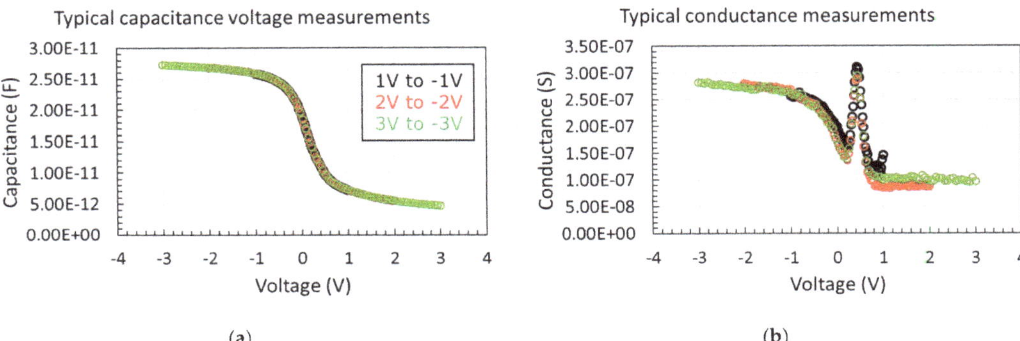

Figure A1. Typical (**a**) capacitance and (**b**) conductance measurements performed in this work. Data shown for Ru/HfO2 combination on a SiO_2 bevel (slant etch) on a p-type Si substrate. The capacitors are sequentially measured at different voltage sweep ranges (i) 1 V to −1 V, (ii) 2 V to −2 V, (iii) 3 V to −3 V.

Figure A2 shows the schematic of an automated V_{FB} extraction with a robust and traceable procedure. Test for gate leakage is performed in the measurement routine (not shown) and warnings are issued if any issues are encountered. Only those data with appropriate fit errors are filtered for further analysis. The rest of the analysis follows as discussed in the main article (see page 5 onwards).

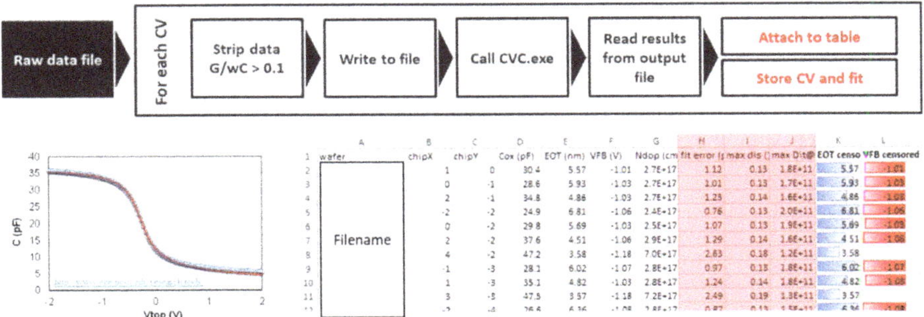

Figure A2. Example procedure of data extraction from measurement.

References

1. Parat, K.; Goda, A. Scaling Trends in NAND Flash. In Proceedings of the IEEE International Electron Devices Meeting (IEDM), San Francisco, CA, USA, 1–5 December 2018; pp. 2.1.1–2.1.4. [CrossRef]
2. Tanaka, H.; Kido, M.; Yahashi, K.; Oomura, M.; Katsumata, R.; Kito, M.; Fukuzumi, Y.; Sato, M.; Nagata, Y.; Matsuoka, Y.; et al. Bit Cost Scalable Technology with Punch and Plug Process for Ultra High Density Flash Memory. In Proceedings of the IEEE Symposium on VLSI Technology, Kyoto, Japan, 12–14 June 2007; pp. 14–15. [CrossRef]

3. Breuil, L.; El Hajjam, G.K.; Ramesh, S.; Ajaykumar, A.; Arreghini, A.; Zhang, L.; Sebaai, F.; Nyns, L.; Raymaekers, T.; Rosmeulen, M.; et al. Integration of Ruthenium-based Wordline in a 3-D NAND Memory Devices. In Proceedings of the IEEE International Memory Workshop (IMW), Dresden, Germany, 17–20 May 2020; pp. 1–4. [CrossRef]
4. Jeon, S.; Han, J.; Lee, J.; Choi, S.; Hwang, H.; Kim, C. High Work-Function Metal Gate and High-kappaDielectrics for Charge Trap Flash Memory Device Applications. *IEEE Trans. Electron Devices* **2005**, *52*, 2654–2659. [CrossRef]
5. Tan, C.-L.; Lavizzari, S.; Blomme, P.; Breuil, L.; Vecchio, G.; Sebaai, F.; Paraschiv, V.; Tao, Z.; Schepers, B.; Nyns, L.; et al. In Depth Analysis of 3D NAND Enablers in Gate Stack Integration and Demonstration in 3D Devices. In Proceedings of the IEEE International Memory Workshop (IMW), Monterey, CA, USA, 14–17 May 2017; pp. 1–4. [CrossRef]
6. Arreghini, A.; Van den Bosch, G.; Kar, G.S.; Van Houdt, J. Ultimate Scaling Projection of Cylindrical 3D SONOS Devices. In Proceedings of the 2012 4th IEEE International Memory Workshop, Milan, Italy, 20–23 May 2012; pp. 1–4. [CrossRef]
7. Charbonnier, M.; Mitard, J.; Leroux, C.; Ghibaudo, G.; Cosnier, V.; Besson, P.; Martin, F.; Reimbold, G. Reliable extraction of metal gate work function by combining two electrical characterization methods. In Proceedings of the ESSDERC 2007—37th European Solid State Device Research Conference, Munich, Germany, 11–13 September 2007; pp. 275–278. [CrossRef]
8. O'Sullivan, B.; Kaushik, V.; Ragnarsson, L.-A.; Onsia, B.; Van Hoornick, N.; Rohr, E.; DeGendt, S.; Heyns, M. Device performance of transistors with high-/spl kappa/ dielectrics using cross-wafer-scaled interface-layer thickness. *IEEE Electron Device Lett.* **2006**, *27*, 546–548. [CrossRef]
9. Akiyama, K.; Wang, W.; Mizubayashi, W.; Ikeda, M.; Ota, H.; Nabatame, T.; Toriumi, A. Roles of oxygen vacancy in HfO2/ultra-thin SiO_2 gate stacks—Comprehensive understanding of VFB roll-off -. In Proceedings of the 2008 Symposium on VLSI Technology, Honolulu, HI, USA, 17–19 June 2008; pp. 80–81. [CrossRef]
10. Kadoshima, M.; Ogawa, A.; Ota, H.; Ikeda, M.; Takahashi, M.; Satake, H.; Nabatame, T.; Toriumi, A. Two Different Mechanisms for Determining Effective Work Function (fm,eff) on High-k—Physical Understanding and Wider Tunability of fm,eff. In *Digest of Technical Papers, Proceedings of the 2006 Symposium on VLSI Technology, Honolulu, HI, USA, 13–15 June 2006*; IEEE: Piscataway, NJ, USA, 2006; pp. 180–181. [CrossRef]
11. Kamimuta, Y.; Iwamoto, K.; Nunoshige, Y.; Hirano, A.; Mizubayashi, W.; Watanabe, Y.; Migita, S.; Ogawa, A.; Ota, H.; Nabatame, T.; et al. Comprehensive Study of VFB Shift in High-k CMOS—Dipole Formation, Fermi-level Pinning and Oxygen Vacancy Effect. In Proceedings of the IEEE International Electron Devices Meeting, Washington, DC, USA, 10–12 December 2007; pp. 341–344. [CrossRef]
12. Kornblum, L.; Rothschild, J.A.; Kauffmann, Y.; Brener, R.; Eizenberg, M. Band offsets and Fermi level pinning at metal-Al_2O_3 interfaces. *Phys. Rev. B* **2011**, *84*, 15. [CrossRef]
13. Mönch, W. Metal-semiconductor contacts: Electronic properties. *Surf. Sci.* **1994**, *299–300*, 928–944. [CrossRef]
14. Yeo, Y.-C.; King, T.-J.; Hu, C. Metal-dielectric band alignment and its implications for metal gate complementary metal-oxide-semiconductor technology. *J. Appl. Phys.* **2002**, *92*, 7266–7271. [CrossRef]
15. Wen, H.-C.; Majhi, P.; Choi, K.; Park, C.; Alshareef, H.N.; Harris, H.R.; Luan, H.; Niimi, H.; Park, H.-B.; Bersuker, G.; et al. Decoupling the Fermi-level pinning effect and intrinsic limitations on p-type effective work function metal electrodes. *Microelectron. Eng.* **2008**, *85*, 2–8. [CrossRef]
16. Akasaka, Y.; Nakamura, G.; Shiraishi, K.; Umezawa, N.; Yamabe, K.; Ogawa, O.; Lee, M.; Amiaka, T.; Kasuya, T.; Watanabe, H.; et al. Modified Oxygen Vacancy Induced Fermi Level Pinning Model Extendable to P-Metal Pinning. *Jpn. J. Appl. Phys.* **2006**, *45*, L1289–L1292. [CrossRef]
17. Yang, Z.C.; Huang, A.P.; Zheng, X.H.; Xiao, Z.S.; Liu, X.Y.; Zhang, X.W.; Chu, P.K.; Wang, W.W. Fermi-Level Pinning at Metal/High-k Interface Influenced by Electron State Density of Metal Gate. *IEEE Electron Device Lett.* **2010**, *31*, 1101–1103. [CrossRef]
18. Kita, K.; Toriumi, A. Origin of electric dipoles formed at high-k/SiO2 interface. *Appl. Phys. Lett.* **2009**, *94*, 132902. [CrossRef]
19. Bersuker, G.; Park, C.S.; Wen, H.-C.; Choi, K.; Price, J.; Lysaght, P.; Tseng, H.-H.; Sharia, O.; Demkov, A.; Ryan, J.T.; et al. Origin of the Flatband-Voltage Roll-Off Phenomenon in Metal/High- k Gate Stacks. *IEEE Trans. Electron Devices* **2010**, *57*, 2047–2056. [CrossRef]
20. Iwamoto, K.; Ogawa, A.; Kamimuta, Y.; Watanabe, Y.; Mizubayashi, W.; Migita, S.; Morita, Y.; Takahashi, M.; Ito, H.; Ota, H.; et al. Re-examination of Flat-Band Voltage Shift for High-k MOS Devices. In Proceedings of the 2007 IEEE Symposium on VLSI Technology, Kyoto, Japan, 12–14 June 2007; pp. 70–71. [CrossRef]
21. Charbonnier, M.; Leroux, C.; Cosnier, V.; Besson, P.; Martinez, E.; Benedetto, N.; Licitra, C.; Rochat, N.; Gaumer, C.; Kaja, K.; et al. Measurement of Dipoles/Roll-Off /Work Functions by Coupling CV and IPE and Study of Their Dependence on Fabrication Process. *IEEE Trans. Electron Devices* **2010**, *57*, 1809–1819. [CrossRef]
22. Padovani, A.; Kaczer, B.; Pesic, M.; Belmonte, A.; Popovici, M.; Nyns, L.; Linten, D.; Afanas'Ev, V.V.; Shlyakhov, I.; Lee, Y.; et al. A Sensitivity Map-Based Approach to Profile Defects in MIM Capacitors From I-V, C-V, and G-V Measurements. *IEEE Trans. Electron Devices* **2019**, *66*, 1892–1898. [CrossRef]
23. Kaushik, V.; O'Sullivan, B.; Pourtois, G.; Van Hoornick, N.; Delabie, A.; Van Elshocht, S.; Deweerd, W.; Schram, T.; Pantisano, L.; Rohr, E.; et al. Estimation of fixed charge densities in hafnium-silicate gate dielectrics. *IEEE Trans. Electron Devices* **2006**, *53*, 2627–2633. [CrossRef]
24. Jha, R.; Gurganos, J.; Kim, Y.; Choi, R.; Lee, J.; Misra, V. A Capacitance-Based Methodology for Work Function Extraction of Metals on High-kappa. *IEEE Electron Device Lett.* **2004**, *25*, 420–423. [CrossRef]

25. Hauser, J.R. North Carolina State Umiversity's CVC. *Computer Analysis Software*, 1999.
26. Park, J.-Y.; Yun, D.-H.; Kim, S.-Y.; Choi, Y.-K. Suppression of Self-Heating Effects in 3-D V-NAND Flash Memory Using a Plugged Pillar-Shaped Heat Sink. *IEEE Electron Device Lett.* **2019**, *40*, 212–215. [CrossRef]
27. Calzolari, A.; Catellani, A. Controlling the TiN Electrode Work Function at the Atomistic Level: A First Principles Investigation. *IEEE Access* **2020**, *8*, 156308–156313. [CrossRef]
28. Wen, H.-C.; Choi, R.; Brown, G.; BosckeBoscke, T.; Matthews, K.; Harris, H.; Choi, K.; Alshareef, H.N.; Luan, H.; Bersuker, G.; et al. Comparison of effective work function extraction methods using capacitance and current measurement techniques. *IEEE Electron Device Lett.* **2006**, *27*, 598–601. [CrossRef]
29. Park, K.J.; Doub, J.M.; Gougousi, T.; Parsons, G. Microcontact patterning of ruthenium gate electrodes by selective area atomic layer deposition. *Appl. Phys. Lett.* **2005**, *86*, 051903. [CrossRef]
30. Pantisano, L.; Schram, T.; Li, Z.; Lisoni, J.G.; Pourtois, G.; De Gendt, S.; Brunco, D.P.; Akheyar, A.; Afanas'Ev, V.V.; Shamuilia, S.; et al. Ruthenium gate electrodes on SiO_2 and HfO_2: Sensitivity to hydrogen and oxygen ambients. *Appl. Phys. Lett.* **2006**, *88*, 243514. [CrossRef]
31. Afanas'Ev, V.V.; Stesmans, A. Internal photoemission at interfaces of high-κ insulators with semiconductors and metals. *J. Appl. Phys.* **2007**, *102*, 81301. [CrossRef]
32. Ramesh, S.; Ajaykumar, A.; Bastos, J.; Breuil, L.; Arreghini, A.; Nyns, L.; Soulié, J.-P.; Ragnarsson, L.-Å.; Schleicher, F.; Jossart, N.; et al. Erase Behavior of Charge Trap Flash Memory Devices using High-k Dielectric as Blocking Oxide Liner. In Proceedings of the IEEE Semiconductor Interface Specialists Conference, San Diego, CA, USA, 6–8 December 2020.
33. Chou, A.I.; Lai, K.; Kumar, K.; Chowdhury, P.; Lee, J.C. Modeling of stress-induced leakage current in ultrathin oxides with the trap-assisted tunneling mechanism. *Appl. Phys. Lett.* **1997**, *70*, 3407–3409. [CrossRef]
34. Iwamoto, K.; Kamimuta, Y.; Ogawa, A.; Watanabe, Y.; Migita, S.; Mizubayashi, W.; Morita, Y.; Takahashi, M.; Ota, H.; Nabatame, T.; et al. Experimental evidence for the flatband voltage shift of high-k metal-oxide-semiconductor devices due to the dipole formation at the high-k/SiO2 interface. *Appl. Phys. Lett.* **2008**, *92*, 132907. [CrossRef]
35. Suarez-Segovia, C.; Caubet, P.; Joseph, V.; Gourhant, O.; Romano, G.; Domengie, F.; Ghibaudo, G. Effective Work Function Shift Induced by TiN Sacrificial Metal Gates as a Function of Their Thickness and Composition in 14 nm NMOS devices. In Proceedings of the International Conference on Solid State Devices and Materials, Ibaraki, Japan, 8–11 September 2014. [CrossRef]
36. Bersch, E.; Di, M.; Consiglio, S.; Clark, R.D.; Leusink, G.J.; Diebold, A.C. Complete band offset characterization of the $HfO_2/SiO_2/Si$ stack using charge corrected x-ray photoelectron spectroscopy. *J. Appl. Phys.* **2010**, *107*, 043702. [CrossRef]
37. Wang, X.; Han, K.; Wang, W.; Chen, S.; Ma, X.; Chen, D.; Zhang, J.; Du, J.; Xiong, Y.; Huang, A. Physical origin of dipole formation at high-k/SiO_2 interface in metal-oxide-semiconductor device with high-k/metal gate structure. *Appl. Phys. Lett.* **2010**, *96*, 152907. [CrossRef]
38. Sivasubramani, P.; Boscke, T.S.; Huang, J.; Young, C.D.; Kirsch, P.D.; Krishnan, S.A.; Quevedo-Lopez, M.A.; Govindarajan, S.; Ju, B.S.; Harris, H.R.; et al. Dipole Moment Model Explaining nFET Vt Tuning Utilizing La, Sc, Er, and Sr Doped HfSiON Dielectrics. In Proceedings of the 2007 IEEE Symposium on VLSI Technology, Kyoto, Japan, 12–14 June 2007; pp. 68–69. [CrossRef]
39. Zheng, X.H.; Huang, A.P.; Xiao, Z.S.; Yang, Z.C.; Wang, M.; Zhang, X.W.; Wang, W.W.; Chu, P.K. Origin of flat-band voltage sharp roll-off in metal gate/high-k/ultrathin- SiO2/Si p-channel metal-oxide-semiconductor stacks. *Appl. Phys. Lett.* **2010**, *97*, 132908. [CrossRef]
40. Sharia, O.; Demkov, A.A.; Bersuker, G.; Lee, B.H. Theoretical study of the insulator/insulator interface: Band alignment at theSiO_2/HfO_2 junction. *Phys. Rev. B* **2007**, *75*, 035306. [CrossRef]
41. Franco, J.; Wu, Z.; Rzepa, G.; Vandooren, A.; Arimura, H.; Claes, D.; Horiguchi, N.; Collaert, N.; Linten, D.; Grasser, T.; et al. Low Thermal Budget Dual-Dipole Gate Stacks Engineered for Sufficient BTI Reliability in Novel Integration Schemes. In Proceedings of the 2019 Electron Devices Technology and Manufacturing Conference (EDTM), Singapore, 12–15 March 2019; pp. 215–217. [CrossRef]
42. Arimura, H.; Sioncke, S.; Cott, D.; Mitard, J.; Conard, T.; Vanherle, W.; Loo, R.; Favia, P.; Bender, H.; Meersschaut, J.; et al. Ge nFET with high electron mobility and superior PBTI reliability enabled by monolayer-Si surface passivation and La-induced interface dipole formation. In Proceedings of the 2015 IEEE International Electron Devices Meeting (IEDM), Washington, DC, USA, 7–9 December 2015; pp. 21.6.1–21.6.4. [CrossRef]
43. Choi, K.; Wen, H.-C.; Bersuker, G.; Harris, R.; Lee, B.H. Mechanism of flatband voltage roll-off studied with Al2O3 film deposited on terraced oxide. *Appl. Phys. Lett.* **2008**, *93*, 133506. [CrossRef]
44. Yamamoto, Y.; Kita, K.; Kyuno, K.; Toriumi, A. Study of La-Induced Flat Band Voltage Shift in Metal/HfLaOx/SiO2/Si Capacitors. *Jpn. J. Appl. Phys.* **2007**, *46*, 7251–7255. [CrossRef]
45. Arimura, H.; Cott, D.; Loo, R.; Vanherle, W.; Xie, Q.; Tang, F.; Jiang, X.; Franco, J.; Sioncke, S.; Ragnarsson, L.-Å.; et al. Si-passivated Ge nMOS gate stack with low Dit and dipole-induced superior PBTI reliability using 3D-compatible ALD caps and high-pressure anneal. In Proceedings of the 2016 IEEE International Electron Devices Meeting (IEDM), San Francisco, CA, USA, 3–7 December 2016; pp. 33.4.1–33.4.4. [CrossRef]
46. Vais, A.; Franco, J.; Lin, D.; Putcha, V.; Sioncke, S.; Mocuta, A.; Collaert, N.; Thean, A.; De Meyer, K. On the distribution of oxide defect levels in Al_2O_3 and HfO_2 high-k dielectrics deposited on InGaAs metal-oxide-semiconductor devices studied by capacitance-voltage hysteresis. *J. Appl. Phys.* **2017**, *121*, 144504. [CrossRef]

 *micromachines*

MDPI

Article

Temperature Impacts on Endurance and Read Disturbs in Charge-Trap 3D NAND Flash Memories

Fei Chen [1,†], Bo Chen [1,†], Hongzhe Lin [1], Yachen Kong [1], Xin Liu [1], Xuepeng Zhan [1,2] and Jiezhi Chen [1,*]

[1] School of Information Science and Engineering, Shandong University, Qingdao 266237, China; 202032768@mail.sdu.edu.cn (F.C.); 15763534878@163.com (B.C.); 201932552@mail.sdu.edu.cn (H.L.); kongyachen@126.com (Y.K.); 201700121107@mail.sdu.edu.cn (X.L.); zhanxuepeng@sdu.edu.cn (X.Z.)

[2] State Key Laboratory of High-End Server & Storage Technology, Testing and Evaluation Research Department, Jinan 250000, China

* Correspondence: chen.jiezhi@sdu.edu.cn

† These authors contributed equally to this work.

Abstract: Temperature effects should be well considered when designing flash-based memory systems, because they are a fundamental factor that affect both the performance and the reliability of NAND flash memories. In this work, aiming to comprehensively understanding the temperature effects on 3D NAND flash memory, triple-level-cell (TLC) mode charge-trap (CT) 3D NAND flash memory chips were characterized systematically in a wide temperature range (−30~70 °C), by focusing on the raw bit error rate (RBER) degradation during program/erase (P/E) cycling (endurance) and frequent reading (read disturb). It was observed that (1) the program time showed strong dependences on the temperature and P/E cycles, which could be well fitted by the proposed temperature-dependent cycling program time (TCPT) model; (2) RBER could be suppressed at higher temperatures, while its degradation weakly depended on the temperature, indicating that high-temperature operations would not accelerate the memory cells' degradation; (3) read disturbs were much more serious at low temperatures, while it helped to recover a part of RBER at high temperatures.

Keywords: 3D NAND flash memory; temperature; endurance; read disturb

Citation: Chen, F.; Chen, B.; Lin, H.; Kong, Y.; Liu, X.; Zhan, X.; Chen, J. Temperature Impacts on Endurance and Read Disturbs in Charge-Trap 3D NAND Flash Memories. *Micromachines* **2021**, *12*, 1152. https://doi.org/10.3390/mi12101152

Academic Editors: Cristian Zambelli and Rino Micheloni

Received: 30 August 2021
Accepted: 22 September 2021
Published: 25 September 2021

1. Introduction

After a decade of rapid technological developments, 3D NAND flash memory has been widely utilized in various kinds of storage applications, especially in file memory-related products such as laptops and data centers. Due to the fact of its ultra-high bit density, lower bit cost, and better performances as well as reliabilities, 3D NAND flash is substituting its 2D counterpart step by step. For charge-trap (CT) 3D NAND flash memory, the endurance can largely be improved because the effects of the tunneling layer degradations are weak, and the program time can be faster because the effects of inter-cell interference (ICI) are well suppressed with larger cell-to-cell space. Recently, a 3D NAND with more than 170 layers was announced by the NAND flash makers [1,2]; more impressively, quadruple-level-cell (QLC, 4 bits/cell), penta-level-cell (PLC, 5 bits/cell), and even hexa-level-cell (HLC, 6 bits/cell) operation modes have been demonstrated [3,4]. All these fundamental developments as well as design-technology co-optimizations (DTCO) will drive 3D NAND flash to the mainstream non-volatile memories in the near future [5].

In NAND flash-based memory systems, a major issue that affects both the performance and reliability is the temperature. In conventional 2D NAND flash with a floating gate (FG) structure, as operation temperature increases, the raw bit error rate (RBER) will increase, and the degradation will become more serious. Therefore, on the one hand, the temperature monitor and controller are necessary in NAND flash-based memory systems with robust reliabilities, and on the other hand, the temperature dependence can be utilized as an accelerator to build a lifetime prediction model in a short-time using the Arrhenius model [6], which shows high accuracy in 2D FG NAND flash. However, due to the special structures of cells and the arrays in CT 3D NAND flash, the failure mechanisms are much more complex, and the temperature effects are quite different. This can explain why the conventional lifetime prediction model loses accuracy when it is applied to 3D NAND flashes [7]. Accordingly, comprehensive understandings of the temperature impacts on a 3D NAND flash are strongly required.

In this paper, systematic characterizations of the temperature dependences have been conducted on CT-type 3D TLC NAND flash memories from −30~70 °C, with focus on the RBER modulations in P/E degradations and read disturbs. By using the FPGA-based raw NAND chip tester, it was experimentally observed that the program/erase time and the RBER in the P/E cycling and read disturbs were highly dependent on the temperature. However, temperature had negligible impacts on the cells' degradation, indicating that the CT 3D NAND is suitable for work at high temperatures with no need to worry about accelerated degradations.

The main contributions of this paper are as follows:

- We characterized the P/E cycling in CT 3D NAND flash memory from −30~70 °C using the raw NAND chip tester. An effective TCPT model was proposed to simulate the program time changes by the P/E cycles and the temperature;
- We characterized the cross-temperature measurements to study the temperature-dependent degradations, indicating that high-temperature operations will not accelerate the degradation of the memory cells;
- We characterized the measurements of temperature-dependent read disturbs. It showed that read disturb degrades at cold temperature, but it helps to recover a part of RBER at high temperatures. The underlying origins are analyzed in detail.

The rest of the paper is organized as follows: Section 2 introduces the background and related work; Section 3 presents the evaluation setup; Section 4 describes the measured results of P/E cycling; Section 5 shows the temperature-dependent read disturbs; finally, Section 6 concludes this work.

2. Background and Related Work

In 2D NAND flash memory, memory cells use FG to store electrons. However, as scaling the cell size to sub-1X nm, serious issues occur from the large cell-to-cell interference and variations in stored charges in FG. It turns out to be extremely difficult to increase the bit density while also guaranteeing the reliability. Different from 2D NAND flash, 3D NAND flash utilizes stacked storage layers to increase the bit density, which settles the problem of flat area scaling and, thus, the key issue turns to be how many layers can be stacked. The comparisons between the two different NAND flash structures are shown in Figure 1a. In a 3D CT NAND flash cell, besides the core oxide and poly-Si channel, the gate stack contains an oxide tunneling layer, silicon–nitride CT layer, oxide blocking layer, and a control gate (CG).

In NAND flash operations, there includes single-level-cell (SLC, 1 bit/cell), multiple-level-cell (MLC, 2 bits/cell), triple-level-cell (TLC, 3 bits/cell), QLC, PLC, and HLC. For 2D NAND flash, SLC and MLC modes are adopted in accordance with the products' requirements; while for a 3D NAND flash, TLC mode has been widely used because of the well-tuned reliability and high cost–performance ratio. A typical threshold voltage (V_{th}) distribution of TLC in a 3D CT NAND flash is shown in Figure 1b, wherein seven program states from A to G levels can be well distinguished. Only a part of the erase state (ER) can be observed due to the negative V_{th} values. Each word-line (WL) consists of three pages, the most significant bit (MSB) page, the central significant bit (CSB) page, and the least significant bit page (LSB). When reading the data from the chip, error bits occur at the overlapping regions between the neighbor states, such as A(B) to B(A) error bits when reading the CSB page at V2 level. Thus, suppressing overlapping regions or optimizing reading voltages are challenges to designing highly reliable NAND flash memory. In addition, for a 3D CT NAND flash, one special concern is the shared common CT layer between neighboring cells.

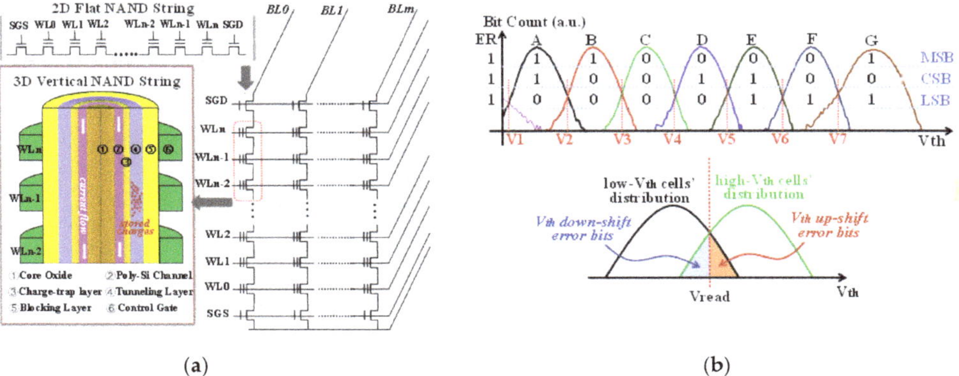

Figure 1. (**a**) A schematic of a 2D NAND string and a 3D NAND string, wherein a 3D NAND cell unit contains core oxide, poly-Si channel, tunneling layer, CT layer, blocking layer, a and control gate; (**b**) TLC operations by storing 3 bits in each cell at three pages: MSB, CSB, and LSB; V1~V7 denote read voltages with the definitions of V_{th} down-shift and up-shift errors.

This makes the reliability mechanisms more complex because the spatial redistributions of stored charges will seriously affect the data retention and read disturbs [8]. It was also reported that the P/E cycling stress affects the charge redistributions more seriously [9]. In other words, in a 2D FG NAND flash, the failure mechanism is simple, and the temperature effects can be monitored, while for a 3D CT NAND flash, the special cell structure makes the failure mechanisms complex and previous temperature-related models are no longer suitable. In the following, several related works are briefly described.

The 3D NAND flash was developed more than ten years ago in 2007 [10], and the first TLC 3D BiCS flash memory with 32 stacked layers was demonstrated by Toshiba in 2015 [11]. Currently, 174 staking storage layers [1,2] as well as HLC operation mode [4] have been realized and demonstrated. So far, 3D NAND flash has been utilized in many kinds of storage products, especially in smartphones, personal computers, and data centers. To assure the robust reliability and high performance of those products, environmental temperature effects should be well considered, and the effects should be included in the system design. Cai et al. studied MLC NAND flash memory and noticed that the error bits from read disturb were much more likely to take place in cells with lower V_{th} values [12]. Zambelli et al. studied cross-temperature effects in 2D and 3D NAND flash memories and found that there was a large number of fail-bits when the memory was read at a temperature different from that exercised during the program [13]. Wu et al. found that cell V_{th} values had various offset and velocities for different temperate operations, and it can be reduced by shortening the interval time from erase to program during cross-temperature write–read stages [14]. Kong et al. studied the read disturbs in a 3D CT NAND flash memory, and observed that read disturbs were strongly correlated to retention time and temperatures, and proposed the schemes of precharge-the-storage-layer (PCSL) and thermally-stabilize-the-storage-layer (TSSL) to suppress read disturbs [8]. Luo et al. observed that the temperature effects increase retention loss speed at a super-linear rate and increases program variations and concluded that prior models for planer 2D NAND flash were not suitable for 3D NAND flash [15]. Resnati et al. investigated the temperature dependence of cell V_{th}, string currents, and random telegraph noise (RTN) distributions in 3D NAND, showing that V_{th} distributions will be tight at high temperatures but widened at low temperatures [16]. In order to minimize the V_{th} distribution widening at low temperature and cross-temperature operations, Venkatesan et al. reviewed the 3D NAND technologies and pointed out that polysilicon channel engineering was necessary [17]. H. Shin et al. investigated the dominant failure mechanisms in 3D NAND after cycling [18] and drew their conclusions that failure mechanisms in 3D NAND are complex; it is not reliable to use temperature as an accelerator for burn-in tests on the basis of the Arrhenius model [7].

As a key factor impacting on NAND flash performance and reliability, temperature effects should be well understood. Thus far, in previous reports, temperature effects have been discussed from material-to-device viewpoints or system-level viewpoints, and most of them focused on data retention. It is necessary to have a comprehensive understanding of the temperature effects to conduct device-to-technology co-optimizations (DTCO) and to provide fundamental information for NAND-based applications in a wide temperature range. In this paper, using a high-performance raw NAND chip tester, temperature-dependent characterizations were performed from −30 to 70 °C by focusing on the RBER in P/E cycling, read disturbs, and cross-temperature degradations.

3. Evaluation Setup

Raw NAND chips were characterized using an FPGA-based raw NAND chip tester with eight parallel sockets and high-speed PCIE interfaces with a maximum data transfer speed up to 200 MHz. The tester was specially designed with a capability to withstand wide temperature operations from −40 to 100 °C, and a customized software was used as the interface to carry out data program/erase/read scripts and detailed data analysis. Moreover, a high–low temperature test chamber with precise controllability to the temperature and humidity was used to perform cross-temperature tests. In our experiments, we chose the 3D CT TLC NAND flash memory chip with 64 stacking storage layers, 5912 valid blocks, with block containing 768 logical pages with 18,336 bytes per page [19]. As shown in Algorithm 1, experiment processes were divided into the following:

- P/E cycling: With combined "Block Program" and "Block Erase" scripts, random data were programmed to the NAND chip and then erased alternately. Here, the generated random data were different in each P/E cycle to ensure the randomness of the characterizations for fair analysis;
- Read Disturb: After data programming, repeated data reading operations were transferred to the tester controller to perform block data reading. The data were not dumped to the controller until we performed a "Block Read Dump" script;
- Data Analysis: The programmed data was read out to the customized software using the "Block Read Dump" operation, and data in the TLC NAND chips were downloaded to a text file. By comparing the programmed data and the read-out data, error bit information could be analyzed.

4. P/E Cycling

The NAND flash memory tested in this experiment was a 3D CT TLC NAND flash chip, and the stored data were divided into eight states according to the V_{th} of the storage cell. Ideally, the V_{th} between adjacent states has a wide read margin, but the V_{th} of adjacent states had overlapping regions in practice, and these overlapping regions were the source of error bits. It should be noted, due to the limited memory window for programming in a high-bit density NAND flash, such as TLC mode, overlapping regions do exist as shown in Figure 1b. The experimental procedure was as follows: firstly, we raised the temperature to the set value; then, we performed 3000 P/E cycles in randomly selected blocks in the chip, and several blocks were characterized to make sure that our results were reliable and stable; finally, the program/erase times of each P/E cycle was recorded in real-time, and the data were exported for analysis.

Algorithm 1. The process of experiment.

```
Definitions:
P_n: the number of P/E cycles, be initialized to 0;
T: the temperature of experiment;
T_target: the target temperature of experiment;
R_count: the number of read cycles, be initialized to 0;
MAX R_n: the maximum number of read cycles;
Process:
if T < T_target or T > T_target then
        Raise T to T_target;
else
        Wait 5 min;
        for P_count ≤ P_n do
                Execute erase/program operation;
                P_count + = 1;
                Collect program time and erase time;
                if P_count == 1 | | P_count % 200 == 0 then
                        Excute the dump operation;
                        Calculate RBER;
                end if
        end for
        Change NAND block;
        While R_count ≤ MAX R_n do
                Excute sequential read operation on NAND block;
                R_count + = 1;
                if R_count == 1 | | R_count % 100 == 0 then
                        Excute the dump operation;
                        Compare and calculate RBER;
                        Collect error classification;
                end if
                end while
end if
```

Figure 2 shows the program time (t_{prog}) and the erase time (t_{erase}) when performing P/E cycling at various temperatures. In the case of t_{erase}, it had a strong dependence on the temperature, but the effects of P/E cycles were negligible. The higher the temperature, the shorter t_{erase}. However, t_{prog} depended on both P/E cycles and the temperature as shown in Figure 2b. For blocks after a certain number of P/E cycles, the higher ambient temperature of the NAND flash memory, the less time it took to execute the program operation. While at the same ambient temperature, t_{prog} decreased with P/E cycles. Thereby, we propose a temperature-dependent cycling program time (TCPT) model on the basis of following equation:

$$t_{prog} = \alpha(T) \cdot n_{pe} + \beta(T) \tag{1}$$

$$\alpha(T) = k_1 \cdot T + k_2,\ \beta(T) = k_3 \cdot T + k_4 \tag{2}$$

T, n_{pe}, and t_{prog} are the temperature, P/E cycles, and program time, respectively. $\alpha(T)$ and $\beta(T)$ are the temperature-related functions with fitting parameters k_n (n = 1~4). As shown in Figure 2b, for blocks with higher than 100 P/E cycles, simulation curves agreed well with the experimental data, indicating that the model was effective at predicting the program latency in a wide temperature range. The values of the fitting parameters are listed in Table 1.

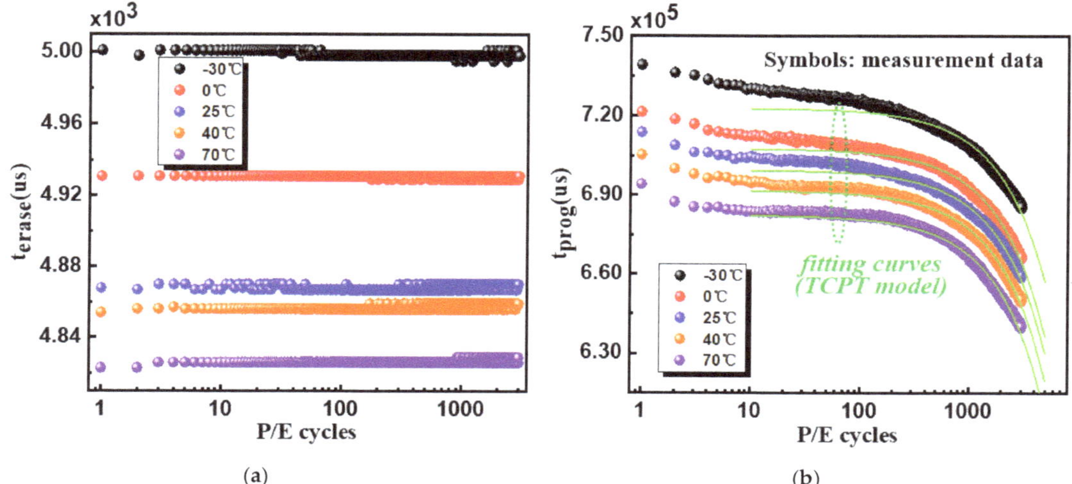

Figure 2. Measured operation times during P/E cycling at different temperatures: (**a**) erase time (t_{erase}) and (**b**) program time (t_{prog}). t_{erase} depends on the temperature, while t_{prog} depends on both the temperature and P/E cycles, which agreed well with the simulation curves in the blocks with higher than 100 P/E cycles.

Table 1. The values of fitting parameters in $\alpha(T)$ and $\beta(T)$.

Fitting Parameter	k_1	k_2	k_3	k_4
$\alpha(T)$	−0.0214	−0.13205	/	/
$\beta(T)$	/	/	−391.98	707913

Figure 3 shows the measured RBER when the P/E cycling was performed at different temperatures. No matter what the temperature we choose, RBER has an initially higher value at the fresh state with a decreasing trend in sub-200 P/E cycles, and then it increases during subsequent P/E cycling. Initial higher RBER could possibly come from the unstable initial stage after factory, and the gradually increased RBER can be explained by P/E cycling stress caused cell degradation. During the P/E cycling, defects generated in both the tunneling layer and CT layer and the stability of the NAND cells became worse. The most important thing was that RBER could be suppressed at high temperatures, while it degraded at cold temperatures. These results can be explained by the V_{th} distributions' changes that depended on the temperature. According to the simulated results by Resnati et al. [16], V_{th} distributions are tight at high temperatures (narrower overlap regions cause fewer error bits) but widened at cold temperatures (wider overlap regions cause more error bits). Furthermore, by normalizing the RBER, it was observed that, although RBER increasing at a rate of 70 °C was greater than that at 0 °C, the degradation tendencies did not show a clear temperature dependence in the whole temperature range as shown in Figure 3b. Considering that different V_{th} distributions had different sensitivities to the cells' degradations, it was necessary to use a unified criterion to study the cells' degradations by cycling stress at different temperatures.

Next, for further evidence, cross-temperature characterizations were conducted to study the temperature effects on P/E cycling stress-related cell degradations. As shown in Figure 4, we designed the following experiment: firstly, we selected three groups of blocks, and all groups performed 1000 P/E cycles at 25 °C (Stage-1); secondly, 1000 P/E cycles were executed in three groups, −30, 25, and 70 °C (Stage-2); then, the temperatures were lowered to 25 °C, and 1000 P/E cycles were subsequently performed on all groups (Stage-3). Finally, each group operated 3000 read cycles at 25 °C. It can be observed that, no matter what the temperature we chose in stage-2, the RBER degradation tendency trends were almost the same in stage-3 and read cycles after cross-temperature P/E cycles. On the one hand, it can be concluded that thermal experiences (up to 70 °C) have negligible impacts on RBER degradation; on the other hand, the 3D NAND is suitable for high-temperature operations because the RBER is lower and the effects of P/E cycling caused damage will not be accelerated at higher temperatures. It should be noted that no matter what the operation mode we adopted, operations at higher temperatures did cause larger degradations to memory cells such as enhanced interface trap generation. Fortunately, these damages do not cause worse error bits degradation using the same criteria (Stage-3) as shown in Figure 4.

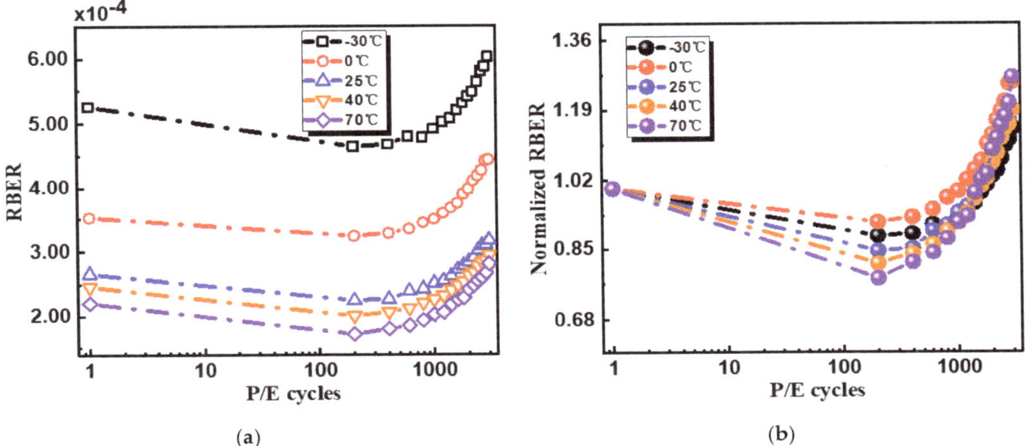

(a) (b)

Figure 3. (**a**) Measured RBER with P/E cycling at different temperatures; (**b**) normalized RBER to study RBER degradation.

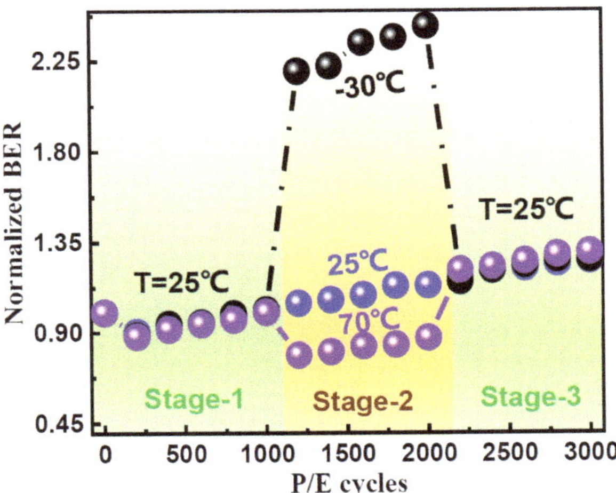

Figure 4. Cross-temperature characterizations to study degradation at various temperatures. The first and third stages were fixed to 25 °C, while the second stage selected three different temperatures, −30, 25, and 70 °C. The RBER was normalized by the first point of the RBER to compare the degradation trends of each condition.

5. Read Disturb

It is known that high temperature can accelerate the speed of lateral charge migration in the storage layer and modulate the threshold voltage distributions of memory cells [8]. These factors can cause read disturb properties to become more complex at various temperatures. To understand the temperature impacts, we designed the following experiment: firstly, setting the work temperature of the chamber to the target temperature ranging from −30 to 70 °C; then, programming randomized data with subsequent 3000 times read cycling. For detail analysis, data were dumped and recorded every 100 times.

Measured read disturbs are summarized in Figure 5. Firstly, it was observed that RBER degradation turned out to be much more serious at cold temperatures; secondly, for high-temperature operations at 70 °C, a part of RBER can be recovered after serval times reading. Considering that the total RBER included two parts, down-shift errors from the charge loss and up-shift errors from charge accrual, we divided the total error bits to two groups for in-depth analysis: down-shift errors and up-shift errors. As shown in Figure 6, for read disturb-related RBER degradations, down-shift errors were the dominant part with clear temperature dependences, indicating that RBER changes mainly originated from the charge loss. Down-shift error degradation was much stronger at sub −25 °C, but it could be well suppressed at high temperatures, which can be explained by the narrower V_{th} distributions at high temperatures [16]. The interesting phenomenon was that read disturbs from up-shift errors showed the opposite tendency while increasing the operating temperature. For read cycling at 70 °C, up-shift error bits can be partly recovered with read cycles. It was noticed that the decreasing error bits were mainly from cells with high program levels, like F-to-G errors in F-level cells. It should be noted that, as shown in Figure 6f, lower F-to-G error bits can be observed in the whole temperature range from −30 to 70 °C. However, G-to-F up-shift error bits are largely suppressed at 70 °C. Thus, with combined down-shift and up-shift errors, we observed abnormal "recovery" at 70 °C while performing read cycling.

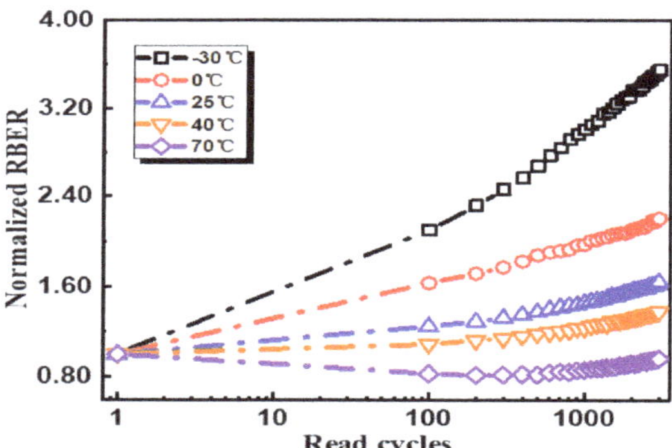

Figure 5. Read disturb characterizations at different temperatures from −30 to 70 °C.

For further understandings, the word-line (WL) dependences of fail bit counts (FBCs) change at 70 °C were studied in detail. By comparing the data from the 1st and 3000th read cycles, it was observed that the dependence of the major state error decreased on the WL index. As shown in Figure 7, error bits from D-E, E-F, and F-G errors showed obvious decreasing trends in higher WL index, and each read cycle in this experiment followed the same observation. In other words, the WLs of the middle-to-low index were the dominant origins for the lower up-shift errors that were attributed to the observed error bits "recovery" at 70 °C.

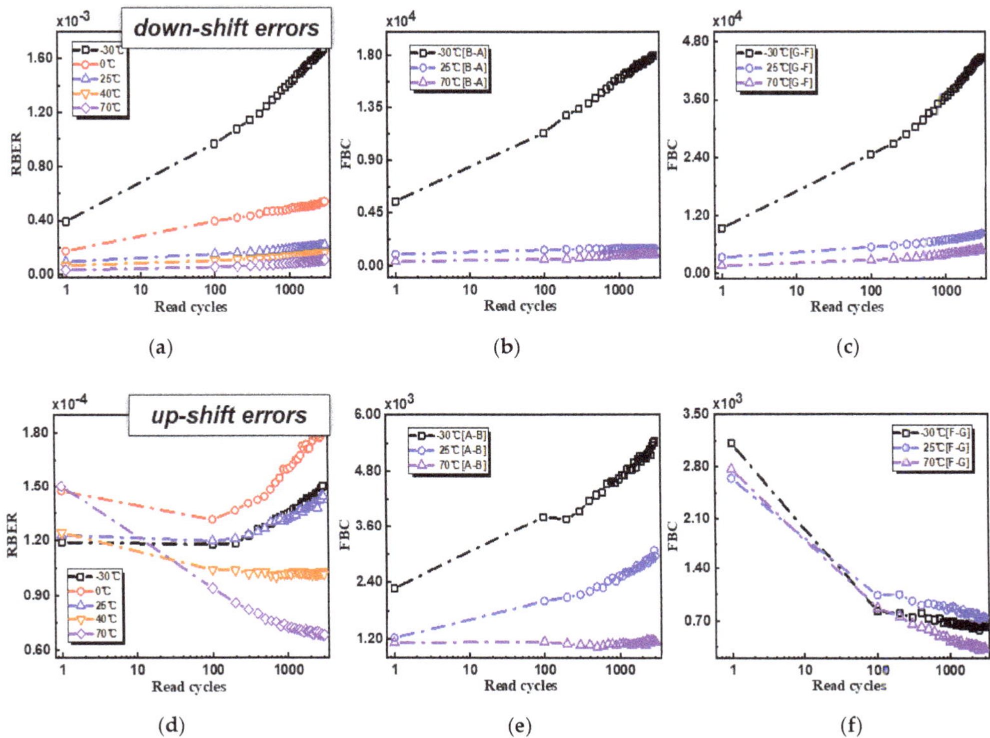

Figure 6. Read disturb-related RBER changes were divided into (**a**) down-shift errors and (**d**) up-shift errors from −30 to 70 °C; (**b**,**c**) compares B-to-A errors and G-to-F down-shift errors, respectively, while (**e**,**f**) compares A-to-B errors and F-to-G up-shift errors, respectively, at −30, 25, and 70 °C.

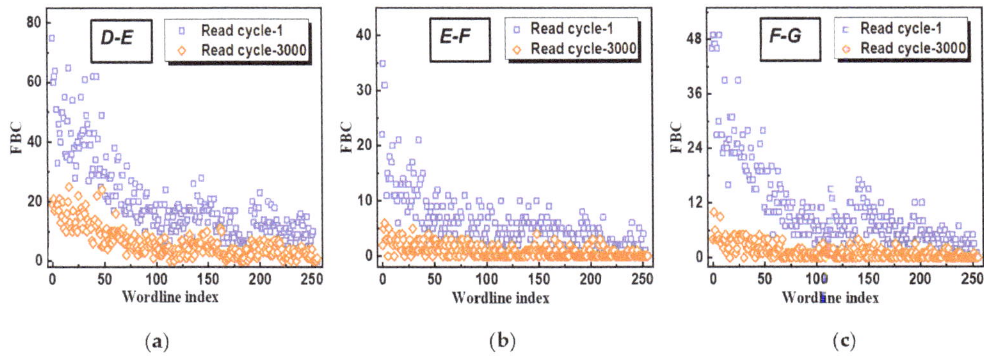

Figure 7. Measured fail bit count (FBC) of different program levels: error bits from (**a**) D-to-E; (**b**) E-to-F; (**c**) F-to-G.

6. Conclusions

In this work, to achieve deep insights into the temperature impacts on the reliability properties of the 3D NAND flash, the TLC (3 bits/cell) 3D CT NAND flash memory chip was tested from −30 to 70 °C using the FPGA-based raw NAND chip tester together with the temperature-controllable chamber. With comprehensive characterizations, firstly, it was observed that program time had a clear dependence on both temperature and P/E cycles by which the TCPT model was proposed; secondly, it was found that RBER can be well suppressed at high temperatures and it degrades obviously at low temperature; then, by the designed cross-temperature measurements, it was found that thermal experience had negligible impacts on RBER degradation; finally, as for read disturbs, it was concluded that read disturbs cause more RBER degradations at cold temperatures while part of RBER can be recovered by read disturbs at high temperatures.

Author Contributions: The work presented here was completed in collaboration between all authors. Conceptualization, F.C. and B.C.; measurements and validation, F.C. and X.L.; simulations, F.C. and H.L.; formal analysis, F.C.; investigation, F.C., B.C., H.L. and Y.K.; writing—original draft preparation, F.C.; writing—review and editing, J.C.; supervision, J.C.; project administration, J.C.; funding acquisition, X.Z. and J.C. All authors have read and agreed to the published version of the manuscript.

Funding: This work was supported by the National Natural Science Foundation of China (No. 62034006 and 91964105), the National Key Research and Development Program of China (No. 2016YFA0201802), and the Natural Science Foundation of Shandong Province (No. ZR2020JQ28 and ZR2019LZH009).

Conflicts of Interest: The authors declare that there is no conflict of interest.

References

1. Park, J.W.; Kim, D.; Ok, S.; Park, J.; Kwon, T.; Lee, H.; Lim, S.; Jung, S.Y.; Choi, H.; Kang, T.; et al. 30.1 A 176-Stacked 512Gb 3b/Cell 3D-NAND Flash with 10.8 Gb/mm^2 Density with a Peripheral Circuit Under Cell Array Architecture. In Proceedings of the 2021 IEEE International Solid-State Circuits Conference (ISSCC), San Francisco, CA, USA, 9 February 2021; Volume 64, pp. 422–423.
2. Cho, J.; Kang, D.C.; Park, J.; Nam, S.W.; Song, J.H.; Jung, B.K.; Lyu, J.; Lee, H.; Kim, W.T.; Jeon, H.; et al. 30.3 A 512Gb 3b/Cell 7th-Generation 3D-NAND Flash Memory with 184MB/s Write Throughput and 2.0Gb/s Interface. In Proceedings of the 2021 IEEE International Solid-State Circuits Conference (ISSCC), San Francisco, CA, USA, 9 February 2021; Volume 64, pp. 426–428.
3. Ishimaru, K. Future of Non-Volatile Memory-From Storage to Computing. In Proceedings of the 2019 IEEE International Electron Devices Meeting (IEDM), San Francisco, CA, USA, 9 December 2019; pp. 1–3.
4. Aiba, Y.; Tanaka, H.; Maeda, T.; Sawa, K.; Kikushima, F.; Miura, M.; Fujisawa, T.; Matsuo, M.; Sanuki, T. Cryogenic Operation of 3D Flash Memory for New Applications and Bit Cost Scaling with 6-Bit per Cell (HLC) and Beyond. In Proceedings of the 2021 5th IEEE Electron Devices Technology & Manufacturing Conference (EDTM), Chengdu, China, 8 April 2021; pp. 1–3.
5. Goda, A. 3-D NAND Technology Achievements and Future Scaling Perspectives. *IEEE Trans. Electron Devices* **2020**, *67*, 1373–1381. [CrossRef]
6. Lee, K.; Kang, M.; Seo, S.; Li, D.H.; Kim, J.; Shin, H. Analysis of Failure Mechanisms and Extraction of Activation Energies (*Ea*) in 21-nm NAND Flash Cells. *IEEE Electron Device Lett.* **2013**, *34*, 48–50. [CrossRef]
7. Zhang, M.; Wu, F.; Yu, Q.; Liu, W.; Wang, Y.; Xie, C. Exploiting Error Characteristic to Optimize Read Voltage for 3-D NAND Flash Memory. *IEEE Trans. Electron Devices* **2020**, *67*, 5490–5496. [CrossRef]
8. Kong, Y.; Zhang, M.; Zhan, X.; Cao, R.; Chen, J. Retention Correlated Read Disturb Errors in 3-D Charge Trap NAND Flash Memory: Observations, Analysis, and Solutions. *IEEE Trans. Comput.-Aided Des. Integr. Circuits Syst.* **2020**, *39*, 4042–4051. [CrossRef]
9. Cao, R.; Wu, J.; Yang, W.; Li, Y.; Chen, J. Error Bit Distributions in Triple-Level Cell Three-Dimensional (3D) NAND Flash Memory. In Proceedings of the 2018 14th IEEE International Conference on Solid-State and Integrated Circuit Technology (ICSICT), Qingdao, China, 31 October 2018; pp. 1–3.
10. Tanaka, H.; Kido, M.; Yahashi, K.; Oomura, M.; Katsumata, R.; Kito, M.; Fukuzumi, Y.; Sato, M.; Nagata, Y.; Matsuoka, Y.; et al. Bit Cost Scalable Technology with Punch and Plug Process for Ultra High Density Flash Memory. In Proceedings of the 2007 IEEE Symposium on VLSI Technology, Tokyo, Japan, 12 June 2007; pp. 14–15.
11. Jeong, W.; Im, J.W.; Kim, D.H.; Nam, S.W.; Shim, D.K.; Choi, M.H.; Yoon, H.J.; Kim, D.H.; Kim, Y.S.; Park, H.W.; et al. A 128 Gb 3 b/cell V-NAND Flash Memory With 1 Gb/s I/O Rate. *IEEE J. Solid-State Circuits* **2016**, *51*, 204–212.
12. Cai, Y.; Luo, Y.; Ghose, S.; Haratsch, E.F.; Mai, K.; Mutlu, O. Read Disturb Errors in MLC NAND Flash Memory. *arXiv* **2018**, arXiv:1805.03283.
13. Zambelli, C.; Crippa, L.; Micheloni, R.; Olivo, P. Cross-Temperature Effects of Program and Read Operations in 2D and 3D NAND Flash Memories. In Proceedings of the 2018 International Integrated Reliability Workshop (IIRW), San Francisco, CA, USA, 7 October 2018; pp. 1–4.
14. Wu, D.; You, H.; Wang, X.; Zhong, S.; Zhong, S.; Sun, Q. Experimental Investigation of Threshold Voltage Temperature Effect During Cross-Temperature Write–Read Operations in 3-D NAND Flash. *IEEE J. Electron Devices Soc.* **2021**, *9*, 22–26. [CrossRef]

15. Luo, Y.; Ghose, S.; Cai, Y.; Haratsch, E.F.; Mutlu, O. HeatWatch: Improving 3D NAND Flash Memory Device Reliability by Exploiting Self-Recovery and Temperature Awareness. In Proceedings of the 2018 IEEE International Symposium on High Performance Computer Architecture (HPCA), Vienna, Austria, 24 January 2018; pp. 504–517.
16. Resnati, D.; Goda, A.; Nicosia, G.; Miccoli, C.; Spinelli, A.S.; Compagnoni, C.M. Temperature Effects in NAND Flash Memories: A Comparison Between 2-D and 3-D Arrays. *IEEE Electron Device Lett.* **2017**, *38*, 461–464. [CrossRef]
17. Venkatesan, S.; Aoulaiche, M. Overview of 3D NAND Technologies and Outlook Invited Paper. In Proceedings of the 2018 Non-Volatile Memory Technology Symposium (NVMTS), Sendai, Japan, 24 October 2018; pp. 1–5.
18. Woo, C.; Kim, S.; Park, J.; Shin, H.; Kim, H.; Choi, G.B.; Seo, M.S.; Noh, K.H. Modeling of Charge Failure Mechanisms during the Short Term Retention Depending on Program/Erase Cycle Counts in 3-D NAND Flash Memories. In Proceedings of the 2020 IEEE International Reliability Physics Symposium (IRPS), Dallas, TX, USA, 27 March 2020; pp. 1–6.
19. Yamashita, R.; Magia, S.; Higuchi, T.; Yoneya, K.; Yamamura, T.; Mizukoshi, H.; Zaitsu, S.; Yamashita, M.; Toyama, S.; Kamae, N.; et al. 11.1 A 512 Gb 3 b/cell flash memory on 64-word-line-layer BiCS technology. In Proceedings of the 2017 IEEE International Solid-State Circuits Conference (ISSCC), San Francisco, CA, USA, 5 February 2017; pp. 196–197.

MDPI
St. Alban-Anlage 66
4052 Basel
Switzerland
Tel. +41 61 683 77 34
Fax +41 61 302 89 18
www.mdpi.com

Micromachines Editorial Office
E-mail: micromachines@mdpi.com
www.mdpi.com/journal/micromachines

www.ingramcontent.com/pod-product-compliance
Lightning Source LLC
LaVergne TN
LVHW070650170726
843515LV00004B/375